Effective Environmental Regulation

Effective Environmental Regulation

LEARNING FROM POLAND'S EXPERIENCE

Halina Szejnwald Brown,
David Angel, and Patrick G. Derr

Westport, Connecticut
London

Library of Congress Cataloging-in-Publication Data

Brown, Halina Szejnwald.
Effective environmental regulation : learning from Poland's experience / by Halina Szejnwald Brown, David Angel, and Patrick G. Derr.
p. cm.
Includes bibliographical references and index.
ISBN 0-275-96971-1 (alk. paper)
1. Environmental policy—Poland. 2. Industries—Environmental aspects—Poland—Case studies. I. Angel, David P. II. Derr, Patrick George, 1949- III. Title.
HC340.3.Z9E5138 2000
363.7'06'09438—dc21 99-098179

British Library Cataloguing in Publication Data is available.

Library of Congress Catalog Card Number: 99-098179
ISBN: 0-275-96971-1

First published in 2000

Praeger Publishers, 88 Post Road West, Westport, CT 06881
An imprint of Greenwood Publishing Group, Inc.
www.praeger.com

Printed in the United States of America

The paper used in this book complies with the Permanent Paper Standard issued by the National Information Standards Organization (Z39.48-1984).

10 9 8 7 6 5 4 3 2 1

Contents

Figures and Tables

FIGURES

TABLES

Acknowledgments

This book represents several years of fruitful collaboration between two institutions: Clark University and the Central Institute for Labor Protection in Warsaw. The vision and high scientific standards of the Institute's Director, Professor Danuta Koradecka, and of the leader of the Polish team, Professor Roman Broszkiewicz, made this project possible and left their mark on the final product. Neither the case studies nor the survey work would have taken place without their leadership, guidance and hard work. They were our partners in designing and implementing the case studies and the survey, and showed an almost magic touch in their ability to open doors to corporate boardrooms, factory floors, corner offices in the Ministries and other leading institutions, as well as the workplaces of many dedicated members of the governmental administration with whom we had a good fortune to speak.

Many other individuals in Poland made invaluable contributions to collecting and interpreting the data for this research. Tomasz Żylicz and Stanisław Wajda of the Ministry of Environment, Andrzej Kassenberg of the Institute for Sustainable Development and, above all, Jerzy Jendrośka of the Environmental Law Center in Wroclaw helped us understand the environmental laws and institutions in Poland, and guided us in asking the right questions. Piotr Gliński of the National Academy of Sciences was the key source of our learning about environmental movements in Poland, past and present. Barbara Krzyśków, Sławomir Wroński and Ewa Suska of the Central Institute for Labor Protection taught us about the occupational protection system in Poland and competently responded to our seemingly endless requests for data, materials and contacts with key individuals to interview. We are immensely grateful to all these individuals. We also express the gratitude to the presidents of the five firms who agreed to become the objects of our case analyses: Eugeniusz Filipczyk of Drumet, Andrzej Wiechno of Fama, Wojciech Topiński of Majewski Pencil Factory, Wiesław Wisniewski of Raffil, and Ewa Matyga, Administrative Director of the Radom Leather Tannery.

At Clark University several graduate students made significant contributions to this project at its different stages: Iain Watt, Rob Krueger, Arvind Susarla, Anna Makri and Olexi Kabyka. Lu Ann Pacenka provided superb assistance in producing the manuscript.

Support for this research came from the National Science Foundation, Program in Ethics and Value Studies, and the International Program, and from the National Council for Eastern European and Soviet Research. We gratefully acknowledge that support.

Chapter 1

Introduction

How can a society organize its system for regulating industrial activity so that sustained gains in environment, health and safety (EH&S) performance are obtained without unnecessarily restraining simultaneous progress in industrial competitiveness and socioeconomic welfare? To what extent can the regulatory approaches adopted by OECD (Organization for Economic Cooperation and Development) nations serve as models for other societies? These questions are particularly important for the rapidly industrializing economies of the developing world, many of which have thus far employed a de facto policy of growth now and cleanup later. In the face of declining air and water quality and mounting local and international pressure there is now growing interest among these industrial economies in strengthening and broadening existing approaches to EH&S protection.

Many of these developing countries have looked to the United States and other OECD economies for possible models of EH&S management. There is certainly much to be gained from studying the OECD experience. But there are at least three compelling reasons to suppose that a broader frame of reference will also be useful. First, regulatory approaches in the OECD countries are themselves undergoing substantial reevaluation and change (OECD 1997). This reevaluation is linked to a growing interest in the roles that civil society, the corporate sector, technological innovation and markets can play as engines for improving EH&S performance. It is also linked to a growing recognition that it is necessary to address EH&S concerns within an analytical and policy framework of sustainable development. And in many cases, it has led to experiments with alternative "softer" regulatory models based on cooperation among parties and flexible policy instruments. Whether (or which of) these new approaches will succeed is still uncertain: regulatory reform may reduce the costs and burdens of regulation (Pearce and Turner 1990; Pearce, Barbier and Markandya 1991; Feiock and Haley 1992) and stimulate greater technological and organizational innovation (Ashford 1991; Heaton 1997; Miller 1995; Brown, Canzler and Knie 1995), but it may also reduce the pressure on firms to improve EH&S performance (Steinzor 1998b). It will likely be a decade before the long-term impact of various reforms can be plausibly evaluated, and few developing countries can afford to defer EH&S regulatory decisions that long.

Second, many aspects of current OECD environmental management approaches may be viewed as too burdensome and too costly to succeed in developing economies already challenged to pursue multiple (and often competing) societal goals (see, for example, Asian Development Bank 1997). This view—whether true or not—is already generating widespread interest among developing economies in alternative regulatory approaches, from market-based instruments to informal regulation and private-law regulatory models. Context-specific approaches that build on the special characteristics of developing economies in general (such as the potentially greater role for industrial and urban policies and the strategic targeting of foreign direct investment and technology transfer) or on the unique characteristics of individual such economies (such as Poland's decisive preference for cooperative rather than confrontational regulatory interventions) need to be considered.

Third, it is not at all clear that we currently know enough to predict which elements of an environmental regulatory system can be expected to function well in a different societal context. Such predictions would require a new and deeper understanding of the relationship between a regulatory system and its economic and societal context. And this kind of understanding can only emerge from the study of a broader array of regulatory experiences, including experiences in Latin America, Eastern and Central Europe, East Asia and elsewhere.

In this book we address the recent experience of Poland in managing the EH&S performance of newly privatized manufacturing firms in a societal context characterized by strong pressure to simultaneously improve EH&S performance and achieve rapid economic development. At first glance the well-documented environmental problems of Poland could plausibly be taken to suggest that there is little to learn from that country's approach to the EH&S performance of industry (French 1990; Institute of Environmental Protection 1990; Kabala 1985; Ember 1990; World Resources Institute 1992). However, three factors suggest that Poland is an especially interesting case for advancing our state of knowledge about EH&S systems. For one thing, after four decades of extraordinary disregard for environmental consequences of rapid industrialization, during the 1990s Poland has made significant objective improvements in EH&S performance (Brown, Angel and Derr 1998; Broszkiewicz, Krzyśków and Brown 1998; GUS 1996a, 1996b, 1998a, 1998b; Cole 1997). Most pollution indices, such as sulfur dioxide emissions and the release of untreated industrial waste, show significant improvement during the 1990s. Occupational fatal accident rates are declining, albeit slowly, and are comparable to those in other OECD countries (World Health Organization 1996). And these improvements have been made while the government has been under enormous pressure to meet other vital societal goals, such as increasing the market competitiveness of industry, maintaining high levels of employment and improving wages and living conditions (Bolan and Bochniarz 1994).

Second, contrary to earlier expectations of Western observers, Poland has made a deliberate effort to pursue improvements in environmental conditions while simultaneously advancing other socioeconomic and political objectives. Despite significant initial economic and social dislocation, Poland has already emerged as one the strongest economies in Central and Eastern Europe, maintaining a high rate of growth of gross domestic product (GDP), industrial production and exports. Real GDP surpassed pre-transition levels in 1994 and grew at an annual rate of 7% in 1997 (OECD 1998). Industrial production, which represents approximately 40% of Poland's GDP, grew at more than 10% in 1997 (GUS 1998a). Unemployment remains high but is falling to close to 10% of the available workforce. A vibrant private sector has emerged and, although barely a decade old, is already responsible for more than 70% of GDP and 60% of employment (World Resources Institute 1998; U.S. Department of Commerce 1998; Ernst 1997). In 1996, 65% of employment in manufacturing was in the private sector, a growth of 6% in one year (GUS 1997, p. 23). Both industrial output and exports by private firms engaged in light manufacturing have been rapidly expanding (Pond 1998; Poznański 1994; Rondinelli and Yurkiewicz 1996). At the same time the environmental quality has improved considerably (see Chapter 2). In sum, despite the dramatic toxic legacies of the communist era, Poland in the 1990s has successfully pursued the goals of environmental protection, occupational health and safety and socioeconomic development.

The third consideration making Poland an outstanding case for studying the simultaneous pursuit of socioeconomic development and environmental and occupational protection is the extent and rapidity of change taking place within the EH&S system. The post-communist regulatory reforms, the shift to a market economy and the political transition to democratic institutions have forced the existing EH&S actors and institutions to systematically reexamine their roles and practices. These multiple simultaneous changes, both within and around the EH&S system, have created a uniquely fertile opportunity to examine how the Polish EH&S system works and what characteristics explain its successes and failures.

An additional analytical opportunity is provided by the fact that the environmental and occupational protection systems in Poland have very different sociopolitical histories and, at least to date, have responded very differently to the post-communist societal transformation (Brown, Angel and Derr 1998; Broszkiewicz, Krzyśków and Brown 1998).

Davies and Mazurek (1998, p. 5) suggest that the effectiveness of a system of pollution control should be judged on two criteria: the adoption of objectives that best advance societal goals and the degree to which these objectives are met. Our research in Poland addresses the second of these two criteria. In examining the success of the EH&S system in Poland, we do not attempt to assess whether the goals of the system are correct. Although this is obviously an important question, it is beyond the scope of this work and perhaps beyond the current competence of North American scholarship. To be sure, the complex set of societal goals which Polish EH&S actors attempt to balance seems very well suited to Poland's circumstances—but whether this is so will best be determined by time and Polish scholarship.

Our work in Poland—consisting of case analyses of five recently privatized enterprises, examination of the systems for environmental and occupational protection as well as their historical development and a survey of approximately one hundred privatized firms—suggests that the EH&S system in Poland is quite effective. By that we mean that it defines explicit expectations and administrative procedures to be met by firms, makes regulatory decisions in a timely manner and without incurring excessive social cost, provides incentives for improved performance, accurately monitors EH&S performance, identifies those firms that are out of compliance with official norms and has the capacity to assess the magnitude of hazards to environment and health and to eliminate high-risk operations.

The apparent success of the EH&S system in Poland may be surprising to some Western observers in view of the recent history of Polish disregard for environmental protection and for enforcing existing laws. In our view the greatest challenge the system faced at the outset of the post-1989 transformation was, in addition to improving the existing policies and administrative apparatus, eliminating the entrenched past practice of disregard for official policies and EH&S objectives. Furthermore, these attitude and behavioral changes would have to take place at the time when many recently privatized enterprises were marginally profitable, hungry for capital and functioning on very short time horizons and when many governmental agencies were understaffed and facing substantial reorganization. In fact, considering the magnitude of the challenge facing the Polish EH&S system in 1989, a reasonable person could have hypothesized a very grim trajectory for the EH&S system: a new entrepreneurial class placing profit ahead of environmental and occupational protection and all the actors, including the enforcement authorities, continuing the old practice of lax enforcement of existing laws. The resulting regulatory system would have been far from effective.

Our research points to several structural characteristics of the EH&S system as primarily responsible for its recent successes. First, it appears that all the key participants—industrial managers, national policy leaders, regional implementing authorities and the technical community—share a commitment to a set of core values and attitudes. These include: regarding the public policies aimed at protection of natural environment and human health and safety as necessary and legitimate; a belief that EH&S decisions require balancing competing objectives for the common good; a decisive preference for negotiation over confrontation in resolving conflicts among competing interests; and a fundamental commitment to due process and the rule of law. These widely shared core values can be traced back to the rich histories of environmental and occupational policies in Poland and to the political culture grounded in centuries of democratic and legal traditions. This wide sharing of values makes it possible for all key EH&S actors to enter into collaborative decision-making mode and to accept solutions initially deemed suboptimal by some or all parties. It also fosters a willingness to seek and accept the opinions of independent EH&S experts, who are trusted by all parties as suitable social mediators.

A second factor underlying Poland's demonstrated ability to pursue EH&S goals without sacrificing socioeconomic welfare is its institutional capacity to make local decisions which are informed by, and sensitive to, the special circumstances of particular cases. This local decision-making capacity has its roots in several factors: Poland's particular policy approach to environmental and occupational protection; an extensive network of self-confident local institutions (inherited from the communist era but also rooted in pre–World War II Polish culture); and a high degree of knowledge on the local level about the regulated enterprises—knowledge which is a product of both the inherited institutional structure and the tradition of negotiation among actors. This capacity for local and case-sensitive decision making allows decision makers to better balance competing societal objectives, reinforces actors' preference for non-confrontational problem solving and helps to build consensus support for particular solutions.

A third strength of the EH&S regulatory system in Poland is its capacity to respond to and learn from its own as well as other countries' experiences. Post-1989 reforms in the environmental regulatory system (and to a lesser degree in the occupational protection system) have selectively adapted elements of OECD systems to Poland's own institutional and regulatory traditions. These selective adaptations were deliberately incremental and maintained the continuity of the existing highly developed institutional structure. The human and institutional ability to make this kind of sophisticated adaptation is the result of a long tradition of policy analysis, debate and innovation that flourished for decades—even under communist rule—among the environmental policy elites in the government and academic sectors. Indeed, the relatively less impressive record of reform and policy innovation in the occupational arena over the past decade can be linked to the much weaker tradition of analysis and debate of occupational issues during the communist era.

The overarching theme which emerges from our findings is the importance of a good "fit" between EH&S institutions and policies on the one hand and their cultural, economic and social context on the other hand. Much of Poland's success in EH&S protection during the 1990s can be traced to the existence of such a fit. By considering ways in which the current EH&S system and its social context may begin to diverge as Poland's societal transition continues, we can also identify future vulnerabilities in the EH&S system.

The empirical findings presented here concerning environmental performance in Poland over the past decade are significant. But the more important contribution of the book lies in advancing in conceptual and analytical terms current understanding of the characteristics of environmental regulatory systems that support sustained improvements in environmental performance in the context of other societal goals. We start by proposing a six-dimension typology that identifies key structural features of a successful regulatory system. Then, using this analytic typology, we explore the significance of political, social and economic context for regulatory success.

EH&S REGULATORY SYSTEMS IN FLUX

Every system of EH&S regulation, whether oriented towards environmental or occupational protection, is a configuration of interacting institutions and policies with a collective mission to further specific societal goals. Our definition of institutions is deliberately broad: it includes the policy-making and enforcement segments of the governmental sector, industrial firms and their representative organizations, organized labor, the public, non-governmental organizations (NGOs) and independent intellectual elites. Our concept of policy is equally broad, including formal laws, regulations and policy instruments as well as less formal but widely accepted decision-making traditions. As these broad definitions imply, we regard EH&S regulation as a dynamic social process requiring interactions among many parties within the boundaries of formal and informal norms.

This definition of an EH&S system is consistent with Davies and Mazurek's conception of a pollution control system, which they define, after Bertalanffy (1968), as "a set of entities that interact with each other and whose [individual or collective] behavior cannot be described or understood apart from the other entities in the system" (Davies and Mazurek 1998; p. 2). This definition has obvious methodological consequences: to study such a system, one must examine a very wide array of processes and participants.

An EH&S regulatory system is successful if it advances EH&S objectives without imposing unreasonable social and economic costs, and does so in ways that enhance rather than undermine the pursuit of other societal goals, such as improvements in socioeconomic welfare and protection of the rights of individuals. The range of societal goals and interests involved in this conception of success is inevitably wide and diverse. But three pervasive themes recur in much recent public debate about EH&S policy: the socioeconomic cost of achieving improvements in industrial EH&S performance; the impacts of EH&S regulation on industrial competitiveness; and the impact of EH&S regulation on economic growth (Bonus and Niebaum 1997; Davis 1992; Morgenstern 1997; Jaffe, Peterson, Portney and Stavins 1995; Porter and van der Linde 1995; Porter 1990; Repetto 1995; Sorsa 1994). Increasingly, policy makers ask whether new approaches to EH&S protection might yield enhanced EH&S performance with lower regulatory expense and less total cost to firms. Can regulatory reform capture so-called "win-win" opportunities by improving EH&S performance *and* enhancing economic competitiveness? This is a central question in current scientific and policy debates over regulatory reform within the OECD (Kopp, Portney and DeWitt 1990; Weale 1995; Jahn 1998; Anonymous 1996; Kosobud 1997; Lotspeich 1998) and also in current discussions regarding the kinds of regulatory approaches that should be adopted by the rapidly industrializing economies of the developing world (Bunyagidj and Greason 1996; Novak 1996; Smith 1994; World Bank 1999).

In the United States, consideration of the social and economic cost of EH&S policy approaches is not new. Since the creation of the Environmental Protection Agency (EPA) and the Occupational Safety and Health Administration (OSHA) in 1970, cost has been a central issue in policy debate and implementation. The legislative basis for considering such costs, however, is uneven in different environmental and occupational statutes. For example, the National Environmental Policy Act (NEPA) of 1969, the Safe Drinking Water Act (SDWA) of 1974, the Federal Insecticide, Fungicide and Rodenticide Act (FIFRA) and the Comprehensive Response, Compensation and Liability Act (CERCLA) of 1980 explicitly direct the EPA to consider economic and other feasibility factors in implementing the environmental prerogatives. The Occupational Safety and Health Act of 1970 also explicitly recognizes the need to balance health with other socioeconomic concerns.

In contrast, the Clean Air Act of 1970 and the Clean Water Act of 1972 are silent on these issues and provide no guidance at all on the question of how or whether ambient quality standards should be balanced against other societal goals. Four decades of U.S. experience suggests that explicit statutory recognition of the need to pursue simultaneously multiple objectives may enhance the likelihood of achieving those objectives. This is perhaps best illustrated by comparing the progress made by the EPA in regulating toxic contaminants in the air under the Clean Air Act to its progress in regulating toxic contaminants in drinking water under the Safe Drinking Water Act. While close to a hundred drinking water standards have been set since the mid 1970s, little regulatory activity has taken place during that time for toxic air pollutants.

The ongoing reassessment of current regulatory practices in the United States and other OECD countries goes well beyond the specificity of legislative goals. The design of institutions and policies, the choice of policy instruments and past assumptions about the roles of the key actors are all being questioned (see, for example, Afsah, Laplante and Wheeler 1996; The Enterprise for the Environment 1998; Steinzor 1998a; National Academy of Public Administration 1996 and 1997; Aspen Institute 1996). In many cases, proposed reforms de-emphasize traditional "command-and-control" and "risk-based" regulatory approaches in favor of pollution prevention, market-based instruments and cooperation among key actors. In part, these proposals are a mid-course correction in response to the perceived shortcomings of existing policies, such as barriers to technological innovation (Heaton, Repetto and Sobin 1991; Environmental Law Institute 1998), inflated costs of social transactions and pollution abatement, suboptimal benefit-to-cost ratios and other inefficiencies (Weber 1998a; Morgernsten and Landy 1997; Davies and Mazurek 1998; Lotspeich 1998; Davis 1992).

But many of these proposals are also a response to significant new developments, such as the increasing availability of good information on EH&S performance (Tietenberg and Wheeler 1998), the globalization of markets and investment, the growing strength of non-governmental organizations and shifting paradigms of business practice (e.g., from pollution control to pollution prevention and from environmental management to industrial ecology and management for sustainability). In particular, much greater attention is now being paid to the role of emerging non-regulatory EH&S performance drivers, such as economic costs savings, consumer preferences for green products, risk assessment by investors and insurers, global best practice standards and concern with issues of sustainable development, including climate change and non-renewable resource depletion.

In response to all these factors, proposals for innovative regulatory reforms have proliferated. These include emissions and use inventories (Seika and Harrison 1996; Cooney 1997; U.S. EPA 1996), environmental performance scorecards (White 1999; Koehler and Chang 1999), negotiated permits, pollution trading (Skea and Sorrell 1999; Farrow and Toman 1999), eco-taxes (Ikwue and Skea 1996), flexible compliance schedules (Fiorino 1995), life cycle assessments (Nieuwlaar and Engelenburg 1996; White, Smith and Warren 1994; MacLean and Lave 1998; DiSimone and Popoff 1997), standardized environmental management systems (Wright 1994), eco-labeling (Potter and Hinnells 1994; Mattoo and Singh 1994; Smith and Potter 1996), clean technology programs (Glasser 1996; Ashford 1991; Schot 1992; Clift 1997), informal regulation (Pargal, Hettige, Singh and Wheeler 1997) and others (see also Allenby and Richard 1994; Davies and Mazurek 1996).

Efforts are also being made to reconceptualize the relationships among the key EH&S actors in terms of partnerships, voluntary agreements (Arora and Cason 1995; Coglianese 1999; Weber 1998a, 1998b; World Bank 2000), codes of conduct (CERES 1998), and international standards, such as ISO 14000 (Marcus and Willig 1997; Voorhees and Woellner 1997; Welch 1998; Wever 1996; Australian Center for Environmental Law 1996). Particular emphasis is given by some to redefining the role of government agencies to include collaborative outreach and technical assistance (Becker and Geiser 1997; John and Mlay 1999). There can be little doubt that these reforms in EH&S regulation are part of a broader transformation of all regulatory structures in response to changes in technology and patterns of industrial development. Clearly, this is an important period of institutional and policy innovation in EH&S regulation.

Many emerging regulatory reform efforts, however, can only be called social experiments that are driven by imprecisely articulated objectives and based more on a knowledge of what has not worked well in the past than on any empirical understanding of what might work better in the future. Both for the purpose of reforming OECD regulatory systems and for the purpose of selecting regulatory models which might have international applicability, we need a deeper and empirically well grounded understanding of how the structural characteristics of EH&S regulatory systems affect their ability to pursue multiple societal needs successfully. We also need a better understanding of how such systems are embedded in their societal context and how they develop and respond to societal change.

ENTRY POINTS FOR STUDYING EH&S REGULATORY SYSTEMS

EH&S regulatory systems have been the object of extensive research over the past three decades in the United States, other OECD countries and internationally. In general such research projects have used one of four different but not mutually exclusive entry points.

The first line of research has endeavored to examine EH&S regulatory performance by examining quantifiable features of the environment or human health. Indicators commonly used in these studies include concentrations of pollutants in environmental media, discharge rates of pollutants into environmental media from mobile and stationary point sources and from non-point sources (U.S. EPA 1996; for the U.S. data see also *National Air Quality Trends Reports* and *State of the Environment*, each produced annually by U.S. EPA and Council on Environmental Quality [CEQ], respectively), biological indicators of ecosystem health, remediation rates for polluted areas (Landsberg et al. 1998; Butterworth 1995; Brown 1993; Spiegel and Yassi 1997; Kjellstrom and Corvalan 1995; Michalos 1997), and others (OECD 1998a, 1998b). Measures of human population health are also used for ubiquitous environmental pollutants for which the dose-response relationship between exposure and human health status is well understood, such as lead or criteria air pollutants. Indirect physical indicators of trends in environmental quality—generally those related to technological processes, such as the use of hazardous substances (Becker and Geiser 1997), energy efficiency and, more recently, a plethora of indicators of industrial environmental performance (Hammond 1996; Turner et al. 1997; Cummings and Cayer 1993; Ditz and Raganathan 1997; Keohler and Chang 1999; White 1999; National Academy of Engineering 1999) are also used in this type of research.

Other studies link physically measurable outcomes to the objectives of specific laws; examples include analyses of the Clean Air Act (Bryner 1995), Clean Water Act (Adler, Landman and Cameron 1993), NEPA (Bartlett 1997; Andrews 1997b; Clark and Canter 1997), Superfund (Davis 1993; Goldman 1994), FIFRA (Wargo 1996), Safe Drinking Water Act and Toxic Substance Control Act (Harrison and Hoberg 1994). Finally, still within this first general line of research, there is a growing body of scholarship that attempts to track overall patterns of energy, materials and resource use, typically with the intent of establishing whether trends in dematerialization and decarbonization of economic activity can be found (see, for example, Costanza, Perrings and Cleveland 1997; Gomułka and Rostowski 1988).

The second general line of research on EH&S regulatory systems has focused on the operation of organizations and institutions concerned with industrial EH&S, including their mutual relationships in a social context. Studies of this type have focused on the adoption of national agendas, on law making, on the interpretation of laws, on the interactions of key players in the political arena, on environmental movements and on institutional structures (McGrew 1993; Barber 1984; Kraft 1997; Dowie 1995; Hajer 1995; Mazmanian and Nienaber 1979; Goggin et al. 1990; Rosenbaum 1997; Sabatier 1993; Kingdon 1995). This line of research is responsible for bringing to light the substantial international differences in regulatory styles (Ottway 1985; O'Riordan 1985; Axelrod 1997; Jasanoff 1987), the strengths and limitations of risk-based and technology-based regulatory approaches (Andrews 1999; Shrader-Frechette 1991), the central role of institutional learning and adaptation (Lee 1993; Smith 1994), and the role of business in pollution management. Research done from this perspective has also focused on the changing role of the industrial sector within national environmental policies and on the expanding national and international environmental agenda in response to the sustainability imperative (Hawken 1993; Shrivastava 1993; Smith 1993; Fischer and Schot 1993; Welford 1996; Welford and Starkey 1996; Roome 1998).

The third general line of research on EH&S regulatory systems has centered on the cost and effectiveness of regulatory approaches. Its primary concern has been the design of policy instruments that can enhance the effectiveness and efficiency of regulatory systems. This line of analysis has become more common in recent years as the command-and-control and risk-based approaches to pollution control have come under increased scrutiny. This perspective has yielded important insights into the utility of economic incentives, the high costs of social conflicts, the importance of technological innovation in pollution prevention, the limits of end-of-pipe pollution control and the consequences of failing to internalize the costs of pollution (Anderson and Leal 1991; Morgenstern and Landy 1997; Tietenberg and Wheeler 1998; Porter 1990; Porter and Linde 1995; Boyd 1998; Farrow and Toman 1999). This perspective is particularly popular among scholars searching for so-called win-win solutions which permit the simultaneous pursuit of economic growth and environmental protection. With reference to developing economies, scholars in this tradition have explored the relationship between environmental regulation and economic development, and in particular, the validity of the hypothesized inverse-U Kuznets curve (according to which environmental quality first falls and then rises with increases in national per capita income) (Selden and Song 1994).

The fourth general line of research on EH&S regulatory systems has viewed EH&S regulation as a social process, an aggregate of many individual decisions that require balancing multiple and often competing values and objectives. Issues of equity, fairness and legitimacy are central to this kind of research. Work in this tradition has contributed to the development of several theoretical frameworks on risk, such as the cultural theory (Rayner 1992; Schwarz and Thompson 1990; Douglas and Wildavsky 1982; Wildavsky 1987; Wildavsky, Ellis and Thompson 1997), the risk perception theory (Slovic 1992; Fischhoff et al. 1978; Covello 1983), the social amplification of risk theory (Kasperson, Renn, Slovic, Emel et al. 1988; Machlis and Rosa 1990), the social arena metaphor (Renn 1992) and others. This kind of work has illuminated the importance of accounting for the values and preferences of all stakeholders and of building trust among them (Stern and Fineberg 1996; Kasperson, Golding and Kasperson 1999). This body of research has also explored the relative strengths of confrontational versus cooperative ways of social decision making, the power of information and the social implications of scientific uncertainty (Freudenberg 1992; Funtowicz and Ravetz 1992; von Winterfeld 1992; Renn, Webler and Wiedenmann 1995a; Andrews 1997a; Shrader-Frechette 1991). The rich critical debate generated by this fourth analytical approach has helped to drive the decentralization movement of the 1990s and has fostered interest in civic environmentalism (John 1994; Rabe 1997; Ringquist 1993; Crowfoot and Wondolleck 1990; Murdock and Sexton 1999). It also irreversibly changed the language of social discourse about environmental regulation: terms such as empowerment, stakeholders participation, local action, risk communication, social trust, iterative risk assessment, and environmental justice have become a part of the public dialogue about environmental issues (Vig 1997; Krimsky and Golding 1992; Gottlieb 1995; Williams 1988; Grant 1997; Yosie and Herbst 1998, Misztal 1996).

Our study of Poland's experience in regulating industrial pollution draws extensively on these four research traditions and attempts to do so in an integrative manner: we examine interactions among all the key EH&S actors, explore the structure and historical origins of existing policies and institutions, reconstruct the implementation decisions made in specific cases and investigate the values and attitudes of the participants. Our approach assumes that EH&S regulation is a dynamic social process occurring in a context of multiple competing societal values and goals.

But while we draw upon all of the approaches noted above, we do so selectively and unevenly. Our main guides for research design, data interpretation and hypothesis formulation are derived from the second and fourth research perspectives. We build on the third line of research—concerned with the cost and effectiveness of regulatory approaches—when we consider the effectiveness of Polish regulatory approaches and policy instruments in achieving EH&S improvements while being sensitive to profound political and social change, economic restructuring and intensified integration into the global economy. We do not, however, attempt to assess the regulatory costs in general or to judge the cost-effectiveness of individual decisions. Our work also makes modest and

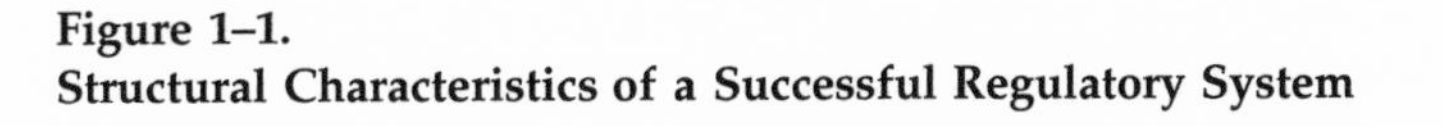

Figure 1–1.
Structural Characteristics of a Successful Regulatory System

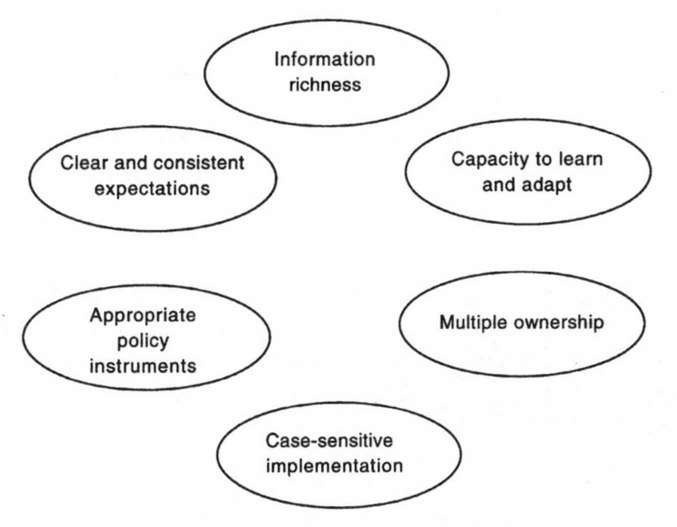

focused use of environmental quality indicators—as these are considered in the first tradition described above—in the context of our case analyses. These data help to illuminate the case histories and the bases and implications of the particular regulatory decisions involved.

At the heart of our study of the Polish experience is an effort to better identify the structural characteristics of a successful EH&S system. The schematic framework for this effort is laid out in Figure 1–1. These are the six structural characteristics of an EH&S regulatory system that in our view are most predictive of success in efforts to realize environmental, health and safety goals without excessive social and economic costs, and in ways that simultaneously achieve desired improvements in socioeconomic welfare. We present these characteristics not as an integrated theory of EH&S regulation, but as a set of linked hypotheses derived both from our prior research and from an analysis of the existing literature.

The first structural characteristic presented in Figure 1–1 is a policy-making process which translates national EH&S policy goals into clear and consistent performance expectations for industry, whether expressed in terms of environmental quality, levels of emissions or other indices. There is a growing body of evidence suggesting that high levels of uncertainty about EH&S performance expectations, or conflicting requirements across media, lead to suboptimal corporate decision making regarding key EH&S investment, technology and management choices (Judge and Douglas 1998; Malone 1990). Regulatory success is most likely when performance standards are clear and there is no ambiguity about regulators' intent to hold industry to those standards. One of the notable features of the Polish system is the ability of local and regional regulators to use their considerable discretionary powers in a way that reinforces, rather than undermines, firms' certainty that they will be held accountable for meeting clearly agreed upon goals.

The availability of appropriate policy instruments is the next structural characteristic presented in Figure 1–1. In our view there is compelling evidence to suggest that the types of policy instruments available to EH&S actors are an important determinant of both cost and performance (Bonus and Niebaum 1997; Luken 1990). The traditional command-and-control regulatory approach relies heavily on permits and mandated technology, typically accompanied by supporting policy instruments such as fines, emission standards and ambient quality standards. The incentive-based approach looks to market-based tools such as full environmental cost pricing, fees, eco-taxes, pollution trading schemes and public disclosure as ways to motivate pollution abatement. With either approach, success requires that the favored policy instruments reliably advance EH&S policy objectives in individual cases without excessive social cost. Of particular importance is the ability to identify and respond to corporate malfeasance in a cost-effective manner. Some scholars have argued that trust, reputation and experience can effectively substitute for more expensive facility monitoring systems and significantly lower overall regulatory costs (Weber 1998a; John and Mlay 1999). The Polish system offers an interesting test case for this hypothesis.

Information richness, another structural characteristic presented in Figure 1–1, is also crucial to effective EH&S regulation. By "information richness" we mean not only general technical knowledge and familiarity with formal and informal EH&S policies and policy instruments but also knowledge about individual corporate facilities, including their technology, management system and performance record. Information richness is less a function of the amount of information available than of the kind and quality of that information. Better performance indicator information, for example, would be extremely useful to all actors involved in EH&S regulation, and we would expect, in the next few decades, a dramatic increase in the quantity and quality of corporate EH&S performance data.

Such data are already being used to track compliance with regulatory standards (Kleindorfer and Orts 1998), and it is clear that EH&S performance data will increasingly be used in multiple other ways, including as a tool of strategic management, as a basis for investment decisions, as criteria for purchasing decisions (Dasgupta, Laplante and Mamingi 1998; Koehler and Chang 1999) and as an instrument for mobilizing public interest and community involvement (Crosby 1999; Tietenberg and

Wheeler 1998). If it is to support these multiple tasks well, performance information will need to be standardized, scaleable, transparently interpretable, and affordable to produce and maintain. Our five case studies found that information richness was central to effective implementation of environmental and occupational policies and that in the Polish context it also included the knowledge and experience that corporate and regulatory professionals have of each other: their shared history and past behavior, their understanding of each others' values and of the external constraints on each others' actions. The case studies also indicate that the process of gathering and interpreting facility-specific information enhances interactions among parties and can facilitate case-sensitive implementation, the next key feature of an effective regulatory system.

Another structural characteristic presented in Figure 1–1 is the ability to make implementation decisions in a way that is sensitive to the specific circumstances and contexts of individual regulatory cases. We suggest that this ability is another key predictor of a regulatory system's success. By "case sensitive implementation" we refer to the capacity of regulators and firms to utilize information about each other within the context of policy implementation. Firms, industries, places and regions vary tremendously in their environmental, economic and social characteristics. Failure or inability to consider such variety leads to inefficient regulation and suboptimal outcomes and will significantly undermine the regulatory system's ability to pursue multiple complex societal objectives. In practice there are many different ways to achieve such case- and context-specific decision making. One view claims that regulatory decision making should be delegated to regional and local authorities. Another view advocates articulating formal rules for negotiated decision making. We treat these claims as hypotheses and believe that the performance of the Polish system, in which decision making with regard to small and medium-size firms is quite responsive to local context, offers a useful test of competing claims about how case- and context-specific decision making can best be achieved within a regulatory system.

The fifth structural characteristic of an effective EH&S system is the capacity to learn and to profit from change. This characteristic involves both an ability to generate and acquire new ideas and scientific knowledge and a willingness to experiment and tolerate uncertainty. It also denotes a capacity for implementing mid-course corrections in policies and policy instruments and a capacity to establish a new regulatory equilibrium in response to major shifts in societal goals and circumstances. This capacity has both external and internal dimensions. A regulatory system must have both the capability to access information generated by external sources and the capability to act on such information. Formal and informal forums for data analysis and policy debate are essential for maintaining an institutional capacity to learn and adapt (Holling 1978; Lee 1993).

The sixth structural characteristic presented in Figure 1–1 is "broad ownership." Key participants in EH&S regulation must have a stake in the success of the regulatory system and must recognize that systemic failure would have consequences inimical to their own interests. (Indeed, a system in which key actors did not regard systemic failure as inimical to their own interests would be heavily handicapped from the outset.) Corporate decisions about compliance with EH&S regulations should be defensible not only in terms of their consequences for the firm but also in terms of their consequences for the system as a whole. In the language of the fourth line of research delineated above, the likelihood that a regulatory system will succeed is strengthened when firms, communities and other groups have "ownership" and are "stakeholders" in the system. When the several institutions comprising the EH&S system have made considerable investments in its development, have considered and rejected possible alternatives and have by their actions demonstrated an active interest in how the system operates and what it produces, the system enjoys multiple ownership. Multiple ownership is a source of creativity and innovation, a buffer against challenges and an effective basis for building information richness and capacity for case-sensitive implementation. Of course, ownership is not equivalent to unquestioning acquiescence by industrial managers to every regulatory decision, nor to disregard by regulators of the corporate or societal impacts of their decisions. However, shared ownership—an acceptance of the fundamental legitimacy of the regulatory process—does put bounds on the disagreement and conflicts which inevitably emerge and fosters a willingness among all parties to seek negotiated solutions.

Collectively, the six structural characteristics of an effective EH&S system comprise the analytical framework which has guided our investigation in Poland. Accordingly, we began our field research with the question: Does the Polish EH&S regulatory system possess these characteristics and, if so, to what extent does their presence or absence explain the perceived strengths and weaknesses of the system?

APPROACH TO THIS STUDY

Our research was conducted in two phases. Phase one involved a careful analysis of Polish institutions, laws and policies concerned with environmental and occupational protection. This work extended to non-governmental institutions such as trade unions, pending legislation and local civic officials responsible for taxation decisions affecting corporate actors. Phase one also included empirical case analyses of five recently privatized industrial facilities. Preliminary findings from this phase of the work were treated as hypotheses for purposes of the project's second stage, a detailed questionnaire survey of 109 privately owned firms.

Our analysis of the institutions, laws and policies sought to elucidate the basic *design and structure* of Poland's emerging EH&S regulatory framework as well as their historical origins. The factory-specific fieldwork focused on the *implementation* of regulatory and corporate policies (see Pressman and Wildavsky 1984 for discussion of implementation). Our inquiry examined how decisions were made at individual industrial facilities and what the outcomes of these decisions were, as assessed in three ways: physical measurements, opinions of the key participants and our own evaluation using the conceptual framework shown in Figure 1-1.

The missions and structures of the institutions responsible for industrial EH&S, the policies they develop and implement and their relationships with other institutions all reflect societal conceptions of environmental and occupational protection in relation to other goods, such as economic growth, employment, national security, individual property rights and societal responsibility—all conditioned by particular cultural, political and historical circumstances. Comparative analyses of EH&S regulatory structures in industrialized countries have underscored the ways in which EH&S regulatory systems are socially constructed. Analysis of the different regulatory responses by European and U.S. societies to environmental and occupational hazards, for example, have highlighted the influence of different cultural traditions (e.g., competition versus cooperation), scientific pluralism, the differing extent to which scientific debate is conducted in the public domain, the relationship between government and the private and public sectors and conceptions of individual versus public rights (Johnson and Covello 1987; Vogel, 1986; Jasanoff 1987 and 1990; Portney 1990; Nelkin 1995; Bryner 1995). In a similar way, our analysis of Polish policies, laws and institutions seeks to elucidate the social context of EH&S regulation in Poland, both in the present and from a historical perspective, and to explicate the background conditions for our case-specific empirical investigations.

Studies combining analysis of regulatory structures with empirical fieldwork within industrial facilities have been relatively rare in environmental and occupational research. Most work has focused either on corporate policies and management systems (Morrison 1991; Kasperson et al. 1988; Smart 1992; Woolard 1992), national EH&S policies and institutions (Portney 1990; Vig and Kraft 1990; Bryner 1995), international comparative policy studies (Jasanoff 1987 and 1990; Johnson and Covello 1987; Vogel 1986) or on specific institutional factors leading to successful or unsuccessful EH&S facility performance (Roberts and Gargano 1990; Rappaport and Flaherty 1991; Roome 1998). Comprehensive analysis of individual facilities in the context of national policies and institutions has for the most part been restricted to retrospective studies of major industrial accidents, such as Bhopal (Shrivastava 1993).

In an earlier study of multinational hazardous facilities in developing countries, we found such a two-pronged methodological approach to be very useful (Brown et al. 1993; see also Durant 1984). Facility-level analysis of the licensing process enabled us to examine the workings of official policies through their implementation and thus to identify conflicts between EH&S-related concerns and other values (whether or not explicitly acknowledged by the principal actors) as these manifested themselves at the local level. Studies of individual facilities also provided a test for hypotheses derived from analyses of regulatory structures and offered new insights for conducting and interpreting such analyses.

In this study we examine both the occupational and environmental protection systems in Poland. Collectively, they are referred to as the system for EH&S management. Traditionally, occupational safety and health and environmental protection have been treated separately, both in the policy-making arena and by scholars conducting research and policy analysis. While the distinction is useful in study design and data analysis, it is instructive to study both systems simultaneously when considering industrial pollution management. Studies of corporate management of pollution and safety have repeatedly found that facility performance in these two areas is linked, partly because they often depend on the same technologies and partly because they both reflect the company's philosophy on such issues as corporate social responsibility, protection of human health, balancing health and safety goals with other corporate objectives, public image and others (Brown, Himmelberger and White 1993; Brown et al. 1993). As we demonstrate later in this book, investigating both the occupational and environmental protection systems in Poland has an additional benefit: a comparative analysis of their relative success raises questions which generate a better understanding of each system.

PREVIEW OF THE BOOK

This book has seven chapters. This introduction, Chapter 1, explains our reasons for selecting Poland as a laboratory for the study of EH&S regulatory approaches, articulates the conceptual framework used in our work and previews the material to follow.

Chapters 2 and 3 describe Poland's environmental and occupational regulatory structures, respectively. These two chapters are both descriptive and evaluative: based on published and unpublished government documents, analytical literature and our own extensive interviews with policy leaders, enforcement personnel, industrial managers, union officials and Polish scholars, they consider current policies and institutions for EH&S in their historical and social context.

Chapters 2 and 3 set the stage for our case studies by highlighting the structural elements of the environmental and occupational protection systems whose performance is assessed by studying the implementation process. They also interpret the current systems as products of three developmental stages in Polish history: the social movements for environmental conservation and worker protection in Poland before World War II; the state-driven development of formal institutions and policies during the communist era; and the post-1989 reform programs implemented during the societal transition toward democracy and a free-market economy.

During the communist period both the environmental and occupational protection systems evolved into sets of sophisticated institutions with considerable technical expertise and elaborate laws and policy instruments. In the environmental arena these efforts were propelled by a vibrant coalition of individual scholars, environmental think tanks and activists with support from the political regime. In the occupational area the state—and ultimately the party—was the sole architect and the engine of change, and open debate was neither encouraged nor supported.

Despite these substantial technical and institutional capabilities, environmental protection policies in Poland were poorly enforced, leading to some of the worst environmental degradation in Europe. The enforcement of occupational standards was similarly lax, although fatal accident rates were comparable to OECD countries. The poor EH&S enforcement record can be traced largely to factors outside the regulatory system, such as control of information, pressures applied to individual firms by the centrally planned economy and an imposed development agenda, which subordinated enforcement of EH&S laws to other priorities such as industrial production and full employment. In addition, some elements of the regulatory system itself contributed to the failure: for example, unattainably strict occupational exposure standards, mandatory compensation for work under dangerous conditions, vague statutory language, inflexible enforcement tools and others.

The incremental reforms to the EH&S regulatory systems since 1989 have strengthened enforcement mechanisms, closed loopholes in existing laws, streamlined administrative procedures and introduced a variety of new policy instruments. The environmental policy elites took advantage of the high tolerance of change during the first years of the democratic rule in order to experiment with untested market-based policy instruments and with other risky new approaches. The response of occupational policy leaders was typical of their recent history: slower, more cautious and departing in incremental steps from the well-established approaches. But they, too, sought to adapt the system to the market economy and diminishing role of the state. Overall, however, the fundamental structures of both systems—their philosophies towards balancing competing objectives, institutional missions, regulatory assumptions and their main supporters—have all remained in place. So have the key actors, both the institutions and individuals staffing them. In general, the continuities have been far deeper than the changes.

The Polish regulatory reform program was predicated upon three crucial but untested assumptions: first, the judgment that the disincentives which had undermined the EH&S regulatory system in the past were primarily external to the system and would disappear with the discredited political regime; second, the belief that the laws, institutions and policies inherited from the communist era were fundamentally sound; and third, the assumption that—confronted with both systemic change in the political and economic domain and incremental reforms in EH&S laws and institutions—regulators, workers, non-governmental organizations (NGOs) and industrial managers would in fact reform their behavior in ways that improved EH&S performance. Our case studies examine the validity of these untested assumptions.

Chapter 4 presents the case analyses of five recently privatized manufacturing facilities. It provides a detailed descriptive narrative of the firms' histories, technologies, environmental and occupational hazards and, most importantly, their interactions with regulators to assess and manage these hazards. The five narratives illuminate the roles and attitudes of the key EH&S actors and pay special attention to the process by which the formal policies described in Chapters 2 and 3 are implemented at specific facilities. The cases also illustrate the differences between formal and informal decision-making practices.

The firms studied in Chapter 4 are of medium size and, while subject to significant EH&S regulations, are not the sort of EH&S dinosaur (such as the *Nowa Huta* steelworks) which attracts the personal attention of central government authorities. Rather, these are "average" firms of the kind that constitute the core of the emerging private economy and are handled by regional and local EH&S authorities. Collectively, the cases represent a wide spectrum of technologies, EH&S problems, commercial situations and technological and managerial sophistication. At one end of the spectrum they include a leather tannery using century-old technology and managing a hazardous waste disposal site that ranks among the most dangerous in the country. At the other end is a modern steel cable manufacturer proudly displaying ISO 9000 and 14000 certificates.

Chapter 5 discusses the major themes which emerge from our analyses of the EH&S regulatory structures and of the case studies. In particular, Chapter 5 highlights the incrementality of policy and institutional reforms since 1989 and the concomitant continuity of personnel in regulatory institutions, firms and policy leadership positions. Chapter 5 also describes the emergence of influential new actors, such as private environmental and occupational health specialists. This chapter suggests that

Poland's EH&S system has been fairly successful in making a transition to democracy and market economy. It demonstrates a considerable capacity for implementing environmental and occupational policies on a facility level while simultaneously balancing multiple competing societal objectives—and typically with little interest or input by workers, trade unions, the public or environmental NGOs.

This chapter also articulates our findings regarding the attitudes of regulators and industrial managers towards certain fundamental issues, such as the need for environmental and occupational regulations, respect for the rule of law, trust in the decision-making process, preference for negotiation over confrontation and knowledge of each other. Information richness and capacity for case-sensitive implementation, using both formal and informal approaches, are among the outstanding features of the EH&S system in Poland. We suggest that these features, as well as the wide sharing of certain values and attitudes, are important factors in the success of the Polish system. Chapter 5 also discusses the slower evolution and progress of the occupational protection system relative to the environmental system, attributing the difference to their respective histories.

Chapter 6 describes the design and results of our random survey of 109 recently privatized firms. The survey was conducted in order to validate and extend the findings of the case studies and policy analysis. It seemed necessary for three reasons. First, the data collected from the cases and from our analysis of Poland's EH&S policies and institutions is richer in its depiction of the attitudes and practices of regulators than in its depiction of the attitudes and practices of industrial managers. Second, our observations are based on the experience of only a small number of firms and may not be representative of the industry as a whole. Third, a verification was particularly important because the success of the EH&S system in Poland, which we describe in Chapter 5, may come as a surprise to many Western and Polish observers. Several scholars predicted during the early years of the transformation period that the past practices of disregard for the environment and for official laws and policies would be slow to change and would interfere with the emergence of an effective EH&S system in post-1989 Poland. In Chapter 6 we report that the results of the survey are consistent with the findings of policy analysis and the case studies. In particular, the survey confirms the hypothesis that managers accept the objectives and policy instruments of the EH&S protection systems in Poland and generally comply with administrative requirements.

Chapter 7, the conclusion, synthesizes the results of the study and evaluates the effectiveness of the environmental and occupational systems in Poland. More importantly, using the typology proposed in Chapter 1, it explains these findings in the context of Poland's political, social and institutional culture and traditions. It then applies these findings to the ongoing debate about environmental regulatory reform in the United States and other OECD economies.

Chapter 7 begins with an assessment of the EH&S regulatory systems in Poland vis-à-vis the conceptual framework proposed above in this chapter. Examined in these six structural dimensions, Poland's environmental protection system shows varied but high degrees of strength on all counts, such as clarity of message about performance expectations by industry; the availability of effective policy instruments; information richness; ability to make case- and context-specific decisions; the sense of ownership of the system among the regulators and industrial managers alike; and a capacity to learn and profit from change. Building upon existing policies and institutions, and after a decade of reforms, environmental regulation in Poland is now a reasonably effective societal institution, with a capacity to pursue EH&S objectives without undue social cost and without interfering with the pursuit of other societal goals. This has been achieved in Poland despite the visible absence of NGOs and other forms of civic engagement in policy implementation.

The occupational protection system, in comparison, has a mixed record. The absence of specific procedures for balancing health and safety against other concerns renders the occupational protection system vulnerable to abuse, while its strengths (such as well codified health and safety standards) are constrained by a low capacity to learn and profit from change and by limited ownership across societal actors and administrative hierarchies. On the other hand, the system is information rich, exhibits capacity for case-specific decisions and is quite effective in articulating clear expectations and responsibilities. Overall, the system's long-term capacity for pursuing its key objectives without undermining other societal goals remains to be tested.

Three factors appear to be important for explaining Poland's relative success over the last decade in developing the institutional characteristics that contribute to effective EH&S regulation: a high degree of positive continuity in institutions, policies and policy instruments within Poland's EH&S system, before and after reform; wide sharing of core values and attitudes among the key societal actors; and broad support for the rule of law and due process. The first, continuity, means maintaining the fundamental structure of the institutions and policies, low personnel turnover and preservation of the accepted mode of conducting societal transactions. The second, shared values, includes a clear preference for negotiation over confrontation and the high value placed on building and maintaining dependable and predictable working relationships among firms, regulatory authorities and independent experts. Additionally, the three groups most active in EH&S policy making and implementation—industrial managers, regional and local government officials and national policy makers—appear to agree that public policies aimed at protection of environmental and occupational health are necessary and that EH&S decisions require balancing multiple objectives for the common good. We are particularly struck by the way that regulatory authorities and firms employ case-specific implementation to balance multiple societal objectives and reach strategic EH&S implementation choices that are sensitive to the nature of hazards, the level of risk and the economic and technical resources of firms.

All these strengths are vulnerable to future challenges. The system's reliance on shared values among the key actors and on continuity in the preferred mode of conducting societal transactions may become problematic as the current generation of managers and regulators are gradually replaced by a new generation whose values and attitudes were formed during the period of profound social transformation. The question of how well the system could adapt to this possibility, or whether it will need to, remains open.

The third factor explaining Poland's success deals with the rule of law and due process. Our study contradicts numerous earlier predictions of a lingering communist legacy consisting of lack of civic values, passivity, learned helplessness and widespread disregard for official laws and policies. While these findings may be surprising to some, we find the broad support for the rule of law and due process revealed by this study to be quite consistent with Poland's well-developed political culture and long history of constitutional and democratic institutions.

The final section of Chapter 7 returns to the opening questions of this book: How generalizable are these findings to other EH&S systems? To what extent can elements of existing successful regulatory approaches—whether in Poland or in OECD nations—serve as models for other societies? How relevant are our findings to ongoing debates about reforming the EH&S systems in the United States, other OECD countries and elsewhere? On this question our discussion builds on the concepts of *information richness* and *case-specific implementation*. These are the threads that link over six criteria of an effective regulatory system to the elements of effectiveness identified in Poland and to the main themes of the recent national debate in the United States and elsewhere over environmental regulatory reform.

Information is not a neutral commodity but rather the product of a dynamic interaction between human judgment, advocacy, the scientific method and technology, which takes place in the process of its generation. Information is an important policy tool not only because it informs decisions and mobilizes societal reaction but also because it provides opportunities for groups and individuals to interact, reach common ground with respect to policy making and implementation, develop working relationships and build trust. Since such interaction takes place in a specific societal context, the meaning of information richness in policy making and interpretation is itself context-dependent. In Poland information richness fits into an established mode of conducting societal transactions which entails negotiations and strategic balancing of competing objectives in specific decision points. Accordingly, information richness includes the knowledge and personal experience that corporate and regulatory professionals have of each other, and it plays a key role in making case-specific implementation decisions possible. At the same time information richness has a limited effect on mobilizing those groups in Poland that traditionally have not participated in local environmental and occupational decision making. In contrast, in the United States, where the mode of conducting societal transactions is more confrontational and community participation is an established tradition, information richness would play a different role in policy formulation and implementation.

In recent years considerable resources have been expended to strengthen the EH&S regulatory systems of rapidly industrializing economies in the developing world. Typically, this has involved the more or less selective transfer of policies and instruments that have seemed effective within the OECD. The result has been a trend towards global best-practice standards in regulatory policies—a result reinforced by the globalization of corporate EH&S management systems, the harmonization of regulatory practices within the European Union and other major trading blocs, the spread of international performance metrics and management systems and the growing influence of international NGOs in shaping regulatory systems.

One potential consequence of these tendencies is reduced sensitivity to the cultural, political and social contexts within which EH&S regulatory systems operate. The diversity of policy experience and experimentation is also likely to be reduced by these global processes. Our results would support a more eclectic approach to EH&S regulation, and we suggest a model that seeks to match regulatory approach to the capacities and characteristics of particular societies. On a general level the six dimensions of an effective regulatory system that we identify have wide applicability in designing and assessing effective regulatory systems in different social, political and economic circumstance in both advanced and developing economies. The significance of country-specific context, we argue, lies primarily in the ways in which these six dimensions of regulatory practice are interpreted and pursued within individual countries. Here we can expect, and indeed should encourage, variation among countries based on specific circumstance, tradition, constraint and opportunity.

Chapter 2

Environmental Protection in Poland

The grim legacy of environmental degradation in Poland under communist rule has been extensively documented in recent years. Since the regime's collapse, technical and popular publications have reported its toxic legacy in detail: suffocating air in the major cities, levels of particulates and sulfur dioxide (especially in the heavily industrialized southwest) far exceeding those in Western Europe and the United States; widespread erosion of historical structures and monuments; dangerous soil contamination with heavy metals; degradation of groundwater; and rivers too polluted to use for industrial processes and agriculture (Kabala 1985; Ember 1990; Institute of Environmental Protection 1990; Fischhoff 1991; World Resources Institute 1992).

Poland in 1989 faced multiple challenges: to ameliorate the worst physical manifestations of past abuses, to improve an environmental regulatory system that allowed some of the worst pollution in Europe and to dismantle the entrenched practice of general disregard for, and lack of enforcement of, existing standards. Three factors worked in favor of environmental progress: the existence in 1989 of a well-developed framework of laws, policies and institutions for environmental protection; the high national profile environmental issues had achieved during the peak years of political opposition to the communist regime; and intense scrutiny and pressure from the international community, accompanied by modest financial and technical assistance. On the other hand, the ambitious environmental agenda for Poland would be pursued at time of profound political and economic change and with the government under considerable pressure to meet other societal goals, such as improved competitiveness of Polish industry in international markets, maintaining high levels of employment and improving wages and living conditions (Cruz, Munasinghe and Warford 1996; Toman 1993; Toman, Cofaba and Bates 1994; Novak 1996; Karaczuń 1995; Jasiński 1996; Pawłowski and Dudzińska 1994; Bolan and Bochniarz 1994; Balcerowicz 1995; Sachs 1993).

The scope of environmental degradation in Poland during the Soviet era led many Western analysts to assume that throughout this period Poland lacked an adequate legal and institutional framework for environmental protection. For example, Hicks (1996, p. 3) writes that "Poland needed—but did not develop—an effective body of legislation and strong environmental administration to mitigate the damage caused by economic activity and rapid urbanization." These assessments are at least partially false.

Early in this century, Poland wrote pioneering laws dealing with the protection and preservation of natural resources. Owing partly to these initiatives, today Poland has one of the last remaining vestiges of primeval European forest and the only thriving herds of European white bison (Mazurski 1997; Perlez 1997). In the post-war period, and most particularly from the late-1970s onwards, Poland passed a series of progressive and innovative laws, developed detailed policy instruments and within some administrative departments built up considerable technical expertise for environmental protection. Many of these efforts were initiated and promoted—with tacit approval from the state—by leading intellectuals and by the increasingly professionalized and pragmatic upper echelons of the Communist Party bureaucracy in response to the evident deterioration of the natural environment in Poland (Cole 1997; Hirszowicz 1990). For example, a research group on environmental law within the Institute of State and Law of the National Academy of Sciences, created in Poland in 1976, was arguably the first such think tank in the Soviet block, and possibly Europe.

In 1989 Poland represented an environmental paradox. On the one hand, it featured a well-developed set of laws, policies and standards for environmental protection; on the other, it endured some of the worst environmental degradation in Europe. The key to understanding this paradox is the realization that the reasons for failure were largely external to the environmental protection system, although the vagueness of the laws, lack of resources for monitoring and enforcement and ineffective administrative apparatus bore some of the blame (Jendrośka 1996a, 1996b; Hicks 1996). The centrally planned economy laid down as non-negotiable such realities as: emphasis on development of heavy industry and mining; heavy reliance on the abundant and cheap but highly polluting coal; a web of personal and economic disincentives for industrial managers to invest into pollution prevention; and a general shortage of capital for environmental protection (Cole 1995a, 1995b; French 1990;

Schmidt and Schnitzer 1993; Toman 1993; Hubbell and Selden 1994; Cole 1997, Kempa 1997; Waller and Millard 1992; Zechenter 1993). These political and economic factors all sharply subordinated environmental protection to industrial production, full employment and economic growth. Moreover, the state's control of information effectively concealed the true scope of the unfolding environmental disaster from all but a relatively tiny elite in the political, administrative and academic sectors (Cole 1997; Gliński 1996, 1998).

This chapter analyses the emergent post-1989 environmental regulatory structure in Poland in the context of its evolution during the twentieth century. It emphasizes the roles and mutual relationships among the key actors, the institutional missions and values, the direction and underlying philosophy of the recent reforms and the links between the current system and its pre-1989 legacy. The analysis is based on extensive examination of policies and institutions, interviews with high level officials and selected physical indicators of environmental performance.

LEGISLATIVE AND INSTITUTIONAL HISTORY BEFORE 1989

The development of environmental legislation in Poland took place in three broad phases. The earliest environmental law focused on preservation of natural resources as the principal approach towards environmental protection and had its roots in Poles' deeply seated respect for nature (Mazurski 1997; Graham 1995). As noted by Mazurski, Poland's pride in its natural resources stretches back to the eleventh century when the first king passed a decree to protect the nation's beavers. The modern era of conservation movement was marked in Poland by legislation implemented in the pre–World War II period, including the 1922 Water Law Act, the creation in 1925 of the State Council for the Protection of Nature and the 1934 Nature Protection Act (Jendrośka and Sommer 1994; Jendrośka 1996b; Cole 1995b, 1997). Cole (1997) and Graham (1995) note that the State Council greatly influenced the history of nature protection in pre-war Poland. Owing to its efforts six national parks and 180 nature reserves were created. Under its influence the subject of nature conservation was incorporated into public school curriculum, and its original publication "Nature Conservation" (*Ochrona Przyrody*) is still published today.

The second phase began with communist rule in Poland and lasted until the late-1980s. During this period Poland was to exhibit a tendency to pass quite progressive and innovative environmental laws but also to fail systematically to implement the laws. The third phase, from 1989 onwards, is marked by economic restructuring, regulatory reform and above all by strengthened enforcement of environmental legislation.

Much has been written about the failure of the communist government to prevent progressive environmental degradation in Poland during the period of communist rule (Kabala 1985; Ember 1990; Institute of Environmental Protection 1990; Fischhoff 1991; World Resources Institute 1992). But the failure was not due to legislative neglect.

After the communists took power in Poland the parliament approved the 1949 Nature Conservation Act, the first of many new environmental laws. This act emphasized "rational use" of natural resources and the environment. In many ways, the law was highly progressive for its times, featuring several provisions that, three decades later, would become the landmarks of the National Environmental Policy Act in the United States: it established a national environmental policy, articulated its goals, established a central agency to carry out its provisions and created a process for evaluating the environmental impacts of various economic activities (not unlike the idea of Environmental Impact Assessment introduced in the United States in 1969). Alongside of policies designed to stem the pollution stream, efforts were also made to protect areas of land from development. Continuing the tradition established during the prewar era, numerous national parks and reserves were created in Poland.

Over the next three decades the Polish parliament passed a plethora of environmental laws and administrative initiatives that were broad in scope and sometimes innovative in approach. These included: creating in the 1960s a civil and penal code on pollution; providing regional administrators with the power to implement pollution policy; creating in 1972 the Ministry for Administration, Local Management and Environmental Protection; introducing in 1974 pollution charges for the use and disposal of water (arguably the first system of environmental fees in the world); creating in 1983 a single-mission environmental agency, the Office of Environmental Protection and Water Management, and in 1987 elevating it to the rank of a Ministry; adopting in 1986 the Air Pollution Act, followed by numerous ambient air quality standards; adopting in 1974 the Water Act to protect surface and groundwater; and adding in 1976 two amendments to the Polish Constitution: "The Polish People's Republic ensures the protection of, and rational management of, the environment" (Chapter 1, Article 12, Section 2) and "Citizens . . . have the right to enjoy . . . natural environment and the duty to protect it" (Article 71) (Jendrośka and Sommer 1994; Jendrośka 1996a).

Many of these legislative efforts were initiated by intellectual elites within the National Academy of Sciences and by the top echelons of the state and the Polish Communist Party itself. As noted by Cole (1997), already in 1971 environmental degradation was a key topic on the agenda of the Sixth Party Congress. During the same year, the Polish Academy of Sciences issued a report entitled "Program of Environmental Protection in Poland to the Year 1990." The report was subsequently adopted by the presidium of the Council of Ministers and the politburo of the party's Central Committee. In 1976 the Polish Academy of Sciences founded a research group on environmental law within its Institute of State and Law, the first such think tank on environmental law in the Soviet block, and possibly in Europe. Cole (1997) richly documents how throughout the 1970s and 1980s environmental protection continued to figure prominently on the national agenda of the party, the parliament and the state administration. Hicks (1996) points out that from the point of view of communist ideology, nature protection and the conservation of natural resources were consistent with the official goal of promoting general social good.

This legislative activity during the period of communist rule culminated in the passage in 1980 of the Environmental Protection and Development Act (EPDA), which created the legal foundation for the state's prominent role in pollution prevention and resource management that is in operation today. Compared with environmental legislation existing in the United States at the time, the EPDA was quite progressive. It addressed all environmental media—air, water and marine environment—as well as waste disposal, protection of flora and fauna and protection of green areas within cities. EPDA also extended the system of fees and fines for water use, in operation since 1974, to other environmental media. Under the provisions of EPDA a separate enforcement agency was created, the State Inspectorate for Environmental Protection (PIOS), and *voivodas* (regional administrators) were empowered to enforce the laws through licenses, fees and fines. The act was also progressive with respect to public participation and litigation rights. For example, unlike environmental groups in many Western European countries at the time, Polish NGOs in principle had the right to file public interest lawsuits in civil courts. The act also accorded to civil associations the right to access financial and environmental information about individual enterprises (Jendrośka and Radecki 1991; Veneziano 1997).

For all of its forward features the EPDA also had its weaknesses, including a lack of judicial precision, a shortage of procedural rules and a tendency to "focus on ends while ignoring the means" (Jendrośka 1998, p. 84). With the benefit of hindsight, researchers would subsequently also question the rigid command and control approach to environmental protection in Poland. But in the specific context of communist rule, two weaknesses were especially damaging. First, the EPDA failed to confront the fundamental contradiction between communist Poland's commitment to environmental protection and its simultaneous reliance upon an economic development strategy that was extraordinarily energy, materials and waste intensive. Second, the EPDA failed to secure the incentives, resources and commitment necessary for successful implementation of the system of environmental standards and requirements.

The EPDA, like earlier legislation, was deeply rooted in the state's philosophy of environmental protection as an instrument to satisfy societal needs (Jendrośka and Radecki 1991). The stated mission of the 1980 act was to specify "the principles of protection and rational shaping of the environment that are intended to assure present and future generations a good standard of living and realization of the right to use environmental resources and to preserve their value" (Chapter 1, Article 1). The act defined the rational management of natural resources as "the use [of] resources only for the purpose of societal interests and while considering not only long-term economic considerations but also their value to ecological balance and human welfare" (Chapter 1, Article 2, Section 1). It further defined shaping of the environment as "affecting the environment for the purpose of achieving social and economic objectives, while simultaneously maintaining balance in nature, especially with regard to renewable resources."

In short the law indiscriminately announced a set of competing social objectives without providing mechanisms for balancing them. Given the national political and socioeconomic priorities under communism, this would lead to disastrous consequences for the environment. To begin with, Poland's economy was very energy intensive. Total primary energy use per unit of gross domestic product (GDP) was in the 1980s about twice the average for Western Europe as a whole. It was also highly dependent on highly polluting lignite and hard coal as a primary source of energy. Prüfer (1997) reports that in 1992, 76.8% of Poland's primary energy came from these sources, among the highest levels in the world. The result was massive amounts of air and water emissions.

At the same time the environmental protection system throughout this period of communist rule largely failed to secure effective treatment (let alone prevention) of pollution, waste and emissions. Investment in environmental protection amounted to no more than 0.5% of GDP during the late 1970s and early 1980s (Prüfer 1997). Jendrośka (1996a, 1996b) and Jendrośka and Sommer (1994) point to the lack of judicial precision and guidance for implementation process in the 1980 Environmental Protection Act, which was a statement of values and aspirations, not a blueprint for societal behavior. But the problems of pollution did not derive principally from any inherent unenforceability of the laws. Rather, they resulted from a shortage of resources, incentives and commitment to enforcement. Emissions standards were routinely exceeded and the environmental protection system was characterized by a chronic lack of monitoring and enforcement. While environmental fees were generally paid, the amounts paid were too small to influence industrial behavior. Moreover, since fees were generally based on approved levels of emissions specified in permits, rather than upon actual levels of emissions, they provided no incentive to reduce emissions (Anderson and Fiedor 1997)! As a consequence, Poland's emissions of SO_2, NOx and particulate matter per unit of GDP in 1989 exceeded the average of all OECD countries by factors of 8.3, 4.5, and 2.3, respectively (Adamson et al. 1996). As much as one-third of municipal and industrial sewage in 1989 was discharged untreated into streams and rivers (Nowicki 1997; Zechenter 1993).

How do we explain the seeming contradiction in the state's action towards the environment: simultaneously developing institutions and policies for its protection and yet failing to enforce them while pursuing economic development strategies that placed ever greater burdens on the natural environment? Some authors (see, for example, Kwaśniewski and Watson 1991; Hicks 1996) see the lack of enforcement in Poland purely as an act of political deception on the part of the communist authorities seeking to create an impression of "liberalization" but having no intentions of implementing the official policies. This view appears to be overly harsh. Rather, we believe that the problem was rooted in Poland's economic structure and post-war national development philosophy. Recounting the post-war history of environmental policy making in Poland and pointing to the party's efforts long before the environment became a symbol of political opposition during the 1980s, Cole (1997) argues persuasively that the upper party echelons appear to have been genuinely concerned with the state of the environment and sincere in their efforts to protect it.

The causes of underenforcement were systemic, deriving from the very nature of the planned economy, the social and political priorities of the communist state, and the control of environmental information by the state. Economic explanations of the failure of environmental protection in Poland highlight the emphasis put on heavy industry and mining, the reliance on highly polluting coal as the primary energy source, the inefficiency of production, the lack of economic incentives for managers of state owned enterprises and the general shortage of capital for environmental protection (Cole 1995a, 1995b; French 1990; Schmidt and Schnitzer 1993; Toman 1993; Hubbell and Selden 1994; Cole 1997; Waller and Millard 1992; Prüfer 1997; Adamson et al. 1996). These economic factors were reinforced by two other processes working in tandem: the lack of clear policy mandates in the 1980 EPDA and a shared understanding among all key parties that environmental protection was subordinate to industrial production, full employment and economic growth. Additionally, the state's control of information, which limited knowledge of the extent of the environmental disaster in Poland to a small elite, prevented the emergence of a strong environmental movement until the 1980s (Hubbell and Selden 1994; Hicks 1996; Cole 1997; Gliński 1996).

To some degree, both the ideological leadership of the country and its intellectual elites and technical experts, many of whom were not supportive of the political system in Poland, shared the same values and concerns with respect to the environment and developed a degree of cooperation within the legal and institutional domains. Polish intelligentsia provided the necessary energy for this wide sharing of environmental values by diverse social actors. The intelligentsia emerged as a distinct social class in Eastern Europe and Russia in the second half of the nineteenth century. This educated elite, often marginalized socially and economically under the conditions of foreign domination and embryonic capitalism, aspired to preserve the heritage of the enlightenment and to provide spiritual leadership for the nation. In the early decades of this century, it was a major force in preserving national traditions and national identity, including the flourishing nature preservation movements (Hirszowicz 1990; Graham 1995; Walicki 1990).

During the communist period the ranks of the pre-war intelligentsia, which were severely and strategically diminished by the occupying German forces, were joined by new highly educated classes, generally of humble social origins. Although in political allegiances the "new" intelligentsia were distinct from the "old," both groups shared such values as respect for professionalism, a sense of collective good and mission and acting on principle (Szablowski 1993a, 1993b; Buchowski 1996). The legacy of nature preservation and environmental activism through state-sanctioned channels was also part of the shared heritage of this emerging post-war social class.

By the 1960s the bureaucratic and political leadership in Poland was increasingly recruited from the ranks of the intelligentsia. During the 1970s, the period of Gierek's leadership, Poland experienced a palpable economic revival and improved standard of living, while the communist party showed signs of de-ideologisation. The alignment of the intelligentsia with the state bureaucracy reached its peak during that decade, leading to what Hirszowicz (1990) describes as the "bureaucratization" of the intelligentsia and "professionalization" of the bureaucracy. The real socialism of the 1970s in Poland was a system in which the ruling party was no longer an alien body but a 3-million-strong non-ideological mass organization in which party members and non-party members collaborated in different forms. Szablowski (1993b) and Buchowski (1996) note that the majority of political elite in Poland during the last decade of communism, as well as now, comes from the intelligentsia. The renaissance of the Polish intelligentsia during that period coincides with the flurry of legal and institutional activity in environmental protection. This is not accidental. It would be that very social group that would have high environmental values as well as the technical and legal expertise to structure the foundations of the environmental protection system in Poland.

During the political upheavals of the late 1980s, most of the Polish intelligentsia aligned itself with Solidarity and against the ruling party. By then, the environmental movement had become a powerful political symbol, used by Solidarity and the state alike to seek political legitimacy (see, for example, Hicks 1996; Waller and Millard 1992; Tatur 1995). This high visibility served to further increase the already high popular support for environmental protection measures. It was thus only natural to see environmental professionals strongly represented at the historic 1989 roundtable discussions between the Solidarity and newly elected noncommunist government. Many members of that group, and their peers, subsequently became environmental policy leaders within bureaucratic and academic circles during the post-1989 period.

DIRECTION OF THE POST-1989 REFORMS

Poland entered the post-communist era of environmental protection with numerous strengths: an extensive system of environmental laws, policies and administrative procedures; strong support for environmental protection among the bureaucratic apparatus, professionals, environmental organizations and the general public; high visibility for environmental issues; considerable technical expertise in pollution management; and significant experience in debating environmental issues, policy making and experimentation. It was also burdened with a long legacy of disregard for official policies and, as described later in this chapter, with a cumbersome command-and-control regulatory system that gave little explicit discretion to regional administrators to account for case-specific circumstances in implementation decisions.

The lack of enforcement of environmental standards and regulations also meant that many policies, implementation instruments and institutional practices were largely untested. As a result there was in 1989 only a limited empirical basis on which to assess the strengths and limitations of the regulatory structure inherited from the past and its suitability for an emergent free-market economy.

The incoming Solidarity government moved swiftly to address environmental concerns. Environmental problems were prominently featured at the historic roundtable discussions between the communist government and the Solidarity opposition, resulting in a release of the "Environmental Protocol" (Jendrośka 1998). The protocol included 28 goals to be realized in 1989–1990, including new economic, legal and administrative instruments of environmental protection. In the initial enthusiasm for radical change in all aspects of political life an assumption dominating the roundtable discussions was that that the disastrous state of the environment in Poland was due to some fatal flaw in the design of existing institutions, laws and regulations, and that improving environmental performance would require their complete overhaul. Goal 10, for example, called for enactment of comprehensive environmental protection reform legislation by the end of 1990 (Jendrośka 1998).

The Environmental Law Reform Taskforce, consisting of 70 experts from academia, government and ecological associations, was charged in 1989 with the task of drafting these reforms. The taskforce and the government quickly dropped the idea of radical changes as neither feasible nor advisable. The "White Paper on National Environmental Policy" that was released in May 1991 called instead for a program of incremental amendments to the laws and reforms of institutions and policy instruments (Ministry of Environmental Protection 1991). In short Poland chose to invest in its considerable capital of existing legal framework, policies and institutions.

During the 1990s the environmental protection system in Poland has evolved in several areas: improving the enforcement and creating incentives for compliance; improving the legal and regulatory framework; extending capabilities for environmental monitoring; introducing explicit procedures for balancing environmental protection and development; increasing the use of economic instruments; increasing the flexibility of authorities to negotiate with enterprises; providing for public participation in policy implementation; incorporating principles of sustainable development into environmental protection policies; and moving towards European standards of environmental protection necessary for acceptance into the European Union.

Shortly after the adoption of the roundtable accords, all environmental and nature protection responsibilities were consolidated in a single Ministry of Environmental Protection, Natural Resources and Forestry, which promptly announced a new National Environmental Policy for Poland (Ministry of Environmental Protection 1991; Cole 1995b). In the spirit of public disclosure, one of the first actions of the new government was to issue a list of the 80 worst polluters. Many of these have since closed down (Warner 1996). The ministry and the parliament then moved rapidly to strengthen the enforcement regime by increasing the powers of the independent enforcement agency, State Inspectorate for Environmental Protection (PIOS), by increasing fees for the use of the environment and fines for non-compliance and by codifying the process of issuing operating licenses to new and existing enterprises.

The extensive system of environmental fees and fines has enabled Poland to create a thriving three-tiered National Environmental Protection Fund and Water Management (on the national, voivodship and local levels) and a National Environmental Bank. The Environmental Protection Fund, which provides subsidies and loans to individual enterprises for environmental improvements, is separated from general revenues and governed by independent boards. As of the mid-1990s, environmental fees and fines generated approximately U.S.$500 million in revenues (Anderson and Fiedor 1997). Partly due to the existence of the fund, Poland's expenditures on the environment have been growing relative to the gross domestic products, from 0.7% in 1990 to 1.4% in 1994, and 1.6% in 1996 and 1997 (GUS 1998a, p. 425). In 1994, for instance, the fund provided close to 40% of all investments into environmental protection in Poland, including foreign assistance (which despite early hopes to the contrary represented only 5% of total), state budget, private loans, corporate sources and others (Budnikowski 1992; Sleszyński 1998; Gomułka 1995). The division of the Environmental Fund into three tiers reflects the ongoing effort to decentralize implementation and enforcement authority in Poland. According to Sleszyński (1998), approximately 10%, 54% and 36% of the revenues from fees and fines are allocated to the national, regional and local funds, respectively. The most recent decentralization efforts, including political and administrative redistricting of Poland, are described later in this chapter.

Since 1989 Poland saw the promulgation of a revamped Nature Protection Act, the amendment of the Environmental Protection and Development Act to include much more detailed policies and procedures, reforms in the Environmental Impact Assessment system, new policies regarding waste management and major revisions in the Water Act (Jendrośka 1998; Mikosz 1996). Alternative approaches to the current command-and-control system have been hotly debated and, in some cases, implemented. Among those, increased flexibility in permitting and enforcement and greater reliance on market-based instruments, especially fees and fines, were most successful (Novak 1996). Other approaches, such as multimedia operating permits, voluntary agreements, generally applicable emission limits, tradable pollution schemes and publicly accessible emission inventories, have taken second stage on the reform agenda (Toman et al. 1994; Żylicz 1994; Warner 1996; Bluffstone and Larson 1997; Linpinski and Otto 1996).

Changes intended to provide greater flexibility in the licensing of industrial enterprises and to mobilize local governments to an active role in environmental protection took longer; but these, too, emerged by the end of the first decade of post-communist reforms. Through a new Compliance Schedule Program the licensing authorities were given discretion to negotiate with (usually large) enterprises in some industrial sectors as to the conditions of operating permits. Regional enforcement authorities (PIOS) acquired the power to negotiate stage-wise enforcement schedules with individual enterprises. Permitting procedures for new facilities now also give local governments veto power in some situations over the decisions of the environmental protection bureaucracy as well as formal appeal mechanisms.

Polish governmental officials were surprised, in privatization proceedings, at the level of concern expressed by potential foreign investors regarding environmental liability. In many cases these proved to be insurmountable obstacles to completing transactions. In other cases investors used the specter of liability as an effective bargaining tool in lowering the prices of enterprises, thereby defeating one of the government's main objectives: to enrich the national treasury. The Ministries of Environmental Protection and of Ownership Changes have responded on a case-by-case basis, unfortunately, leaving many lingering problems for the future. Generally, the cost of cleanup is settled through negotiations, subtracted from the purchase price and placed in a special escrow account accessible by the purchaser (Kristiansen 1996; Zechenter 1993; Jasiński 1996; Warner 1996; Bell 1994; Rider and Zajicek 1995; Wajda and Sommer 1994; Goldenman 1995).

Poland has also made an effort to move away from its earlier conception of the environment as a means for satisfying social needs towards the principles of sustainable development. Thus the proposed aim of the currently debated revisions in the Environmental Protection and Development Act is to "create conditions for sustainable development necessary to secure the existence of man and the right quality of the environment, in the interest of future generations" (Chapter 1, Article 1.1). The act would also make a commitment to balancing environmental protection and other societal interests in ways that benefit the "welfare of society as a whole" (Chapter 1, Article 1.3). More specifically, it suggests that the costs imposed on enterprises should be proportional to the environmental benefits achieved (Chapter 1, Article 1.14). Legislatures are actively considering introducing large scale emission trading as well as, to the extent that it is economically and technically feasible, best available technology (BATEEC). Other proposals would lessen the reliance on pollutant-specific and media-specific permits in favor of integrated environmental permits.

In addition to the legal and policy reforms, the 1990s witnessed both the maturation of the ecological movement and a decline in its political power. By one estimate, in the early 1990s there were 300–400 NGOs concerned with environmental issues in Poland (Nowicki 1997). Most were strictly local groups concerned with environmental problems in their immediate area. The maturation process included mergers of smaller groups, coordination of activities, greater emphasis on serving the needs of the general public and an ideological shift towards the mainstream. Environmental organizations have developed links and formal modes of cooperation with other societal institutions, including the state administration, the legislature, local elected governments, the academy, the private sector and others as the movement became increasingly professionalized in technical and organizational areas (Gliński 1996). NGOs and the public at large have achieved some notable successes in influencing regulatory decisions, such as forcing the government in 1993 to revoke rollbacks on pollution fees for large energy producers (Sleszyński 1998). By and large, however, after the heyday of the roundtable talks, the subsequent participation of NGOs in the policy process has been modest.

The seeds of the current political weakness of the ecological movement were planted during the height of the pre-1989 Solidarity movement. As noted by Hicks (1996) and Gliński (1998), the Solidarity leadership never showed a deep commitment to the environmental agenda, using it primarily for political expediency. Gliński identifies several other key factors in the movement's political demise during the 1990s. First, its leadership is experiencing difficulties in making a transition from protesting and challenging existing political structures towards constructive and creative action. Second, the membership is preoccupied with economic survival in the market economy (see also the discussion of the decline of civic organizations in Poland by Bernhard 1996 and Veneziano 1997). Third, large segments of the population, especially youth, are relatively alienated from the political process, viewing it as unethical and deceitful. Fourth, the Solidarity elites have marginalized the environmental organizations by excluding them from the decision-making process. From its inception as the only independent political force in a Soviet state, political power within Solidarity was closely guarded by small elites and decisions were made behind closed doors (Gliński 1998).

The reform program has so far made little progress towards increasing the participation of local and regional authorities and the public in policy making and implementation. Despite the EPDA's strong provisions for public legal challenges, the opportunities for public participation in licensing of new facilities and in Environmental Impact Assessment proceedings (see the next section), and the proposed provision in the currently debated amendments to the EPDA for individuals' right to demand administrative action against polluters, the real impact of these policies is not likely to be significant. This is because of the limited access to information about industrial performance, licensing conditions, emissions from individual facilities and others, and because of bureaucratic aversion to public hearings. The key reason is the reformers own ambivalence about sharing information and their lack of confidence in the ability of citizens and local authorities to make judgments on matters that have traditionally been the province of experts. Jendrośka (1998) attributes these attitudes largely to Poland's entrenched legal and administrative tradition of secrecy and disdain for lay persons' opinions. Lack of experience among all parties in creating an open, participatory process for policy making also plays a role. The failure to incorporate the requirement for public hearings into accepted administrative practices only exacerbates these tendencies. The fact that the amendments to the Environmental Protection and Development Act have been actively debated for over two years with no significant input from practitioners or the public illustrates the insularity of the policy-making process in Poland.

The process of post-1989 reforms in the environmental protection system also drew on a relatively small policy elite within the administration, professionals, environmental organizations and representatives of the largest industrial enterprises, mainly in the energy sector. Representatives of small and medium-size firms and practitioners were notably absent from these discussions. Tymowski (1993) noted a similar tendency to depend on small centralized elites for effecting reforms in other societal domains in Poland and its paradoxical consistency with the inherited, and presumably discredited, approach to policy making before 1989.

The environmental policy elite in Poland has demonstrated over the past three decades that it is a formidable force. It has a good grasp of the strengths and weaknesses of the existing system, a strong vision of the future, a commitment to environmental protection independent of political ideology and a capacity to influence national agenda setting and policy making. However, the lack of consultation with practitioners within industry and the state bureaucracy opens to question the degree to which the industry and the public are "stakeholders" in the emergent system. Furthermore, the shortage of empirical tests of effectiveness of the past policies and institutions makes it harder to anticipate the impacts of the recent reforms on the behavior of industrial managers and bureaucrats.

While the impetus and ideas for the reforms came from the vital community of professionals within the bureaucracy, NGOs and independent professionals, the driving forces for the reforms have changed during the post-1989 decade. During the early 1990s the reform movement was largely driven by the perceived shortcomings of the system or the desire to experiment with approaches that have proven successful in OECD countries. Such was the case with the use of economic instruments, the introduction of greater flexibility in licensing and implementation, and the establishment of the National Environmental Fund. One of the most innovative ventures from that early period has been the debt-for-environment swap arranged in 1991 with the Paris Club. Between 1991 and 1995, the United States, France, Switzerland, Sweden and Finland agreed to forgive close to one-half billion dollars of Poland's considerable foreign debt in favor of investments into environmental projects with international significance (Bochniarz and Bolan 1998; Anderson 1998).

More recently, the desire to harmonize the Polish system with the requirements of European Union has become the dominant engine for policy change. In response to the general directives issued by the European Commission on April 26, 1996 regarding Poland's proposal for inclusion in the Union, the Polish Parliament issued the "National Integration Strategy" in 1997 and in June 1998 the government adopted the "National Preparatory Program for Joining the European Union." Both documents address ecological integration, especially in the areas of harmonization of the environmental protection legislation, long-term investment strategy in the energy sector, waste management and the agricultural and land-use policies (this is discussed at length in Kamieniecki and Kuspak 1998, pp. 98–109). Some of the proposed changes in the Environmental Protection and Development Act discussed elsewhere in this chapter, such as the adoption of the principles of sustainable development and the timid efforts towards greater public participation, reflect the efforts to join the European Union.

In general, the pace of change in environmental policies and institutions in Poland has slowed down by the end of the first decade of transformation, and the system shows signs of maturation. Policy making and implementation is bureaucratized, the range of policy alternatives has narrowed, the access to the Environmental Protection Fund has become politicized and the active policy elite responsible for the major reforms of the early 1990s are more fragmented and risk averse. Partly, this is a natural sequence in a post-revolutionary process of social change. Moreover, the pace of new policy initiatives declines as the existing reforms move into the implementation stage. Another reason for the emerging plateau stems from the highly political climate in which Poland's environmental reforms have originated. The attenuation of the rate of environmental policy change has followed a similar trend in the national political reforms. In the next sections of this chapter, we discuss the status of the emergent environmental protection system in Poland and its future prospects.

INSTITUTIONAL ARRANGEMENTS DURING THE 1989–1998 PERIOD

The institutional arrangements described in this section were in effect until December 31, 1998. As a result of the decentralization of the governance system in Poland, the institutional structure of the environmental protection system changed on January 1, 1999. We outline these changes further in this chapter. For two reasons we describe in this section the pre-1999 regulatory structure. First, the case studies discussed in Chapters 4 and 5 were conducted in 1997 and can be understood only in the context of the institutional framework effective at that time. Second, the new arrangements have not significantly altered the types of decisions faced by regulators and firms alike.

Despite all the changes since 1989, the fundamental approach to regulating pollution in Poland has remained largely unchanged. It is grounded in a coercive system based on ambient and emission standards, operating licenses, limited flexibility for negotiations between enterprises and regulators, and fines for non-compliance. Central authorities wield the policy-making power, while implementation is delegated, with limited discretion, to regional authorities. We now highlight key features of the current environmental protection system.

A Centralized Single-Mission Environmental Agency with Broad Legislative Powers

Figure 2–1 shows the administrative structure for environmental protection in Poland. The administrative functions of environmental protection are carried out in Poland by a single-mission agency, the Ministry of Environmental Protection, Natural Resources and Forestry (hereafter referred to as the Ministry of Environmental Protection), aided by other environmental agencies, such as the Environmental Impact Assessment Commission as well as other ministries. This single mission sets the Ministry of Environmental Protection apart from the two ministries responsible for occupation safety and health: the Ministry of Labor and Social Policy and the Ministry of Health and Welfare, which must divide their respective attentions and resources between occupational safety and health issues and their other major social missions (e.g., unemployment, labor contracts, health care and social welfare) (see Chapter 3).

The ministry is the chief state organ for preventing pollution and regulating the impact of industrial activities on the natural environment. It has broad responsibilities, including the prevention and regulation of industrial pollution, the management of natural resources and preservation of nature (including national parks and wildlife reservations), the protection and management of fisheries and game, the issuance and enforcement of environmental standards and the setting and collecting of fees for emissions and resource use.

The Council of Environmental Protection is the ministry's advisory and consultative body. Its members are appointed by the minister. The State Environment Protection Inspectorate (PIOS) is the ministry's enforcement arm, with the chief inspector serving as deputy minister of environmental protection. Many of the regulatory activities of the Ministry of Environmental Protection, such as facilities licensing, are delegated to regional state administrators, *voivodas*, and are carried out by the 49 voivodship Departments of Environmental Protection (WWOS). With regard to industrial enterprises, the voivodship WWOS issues environmental permits and collects environmental fees (see below). The heads of voivodship Departments of Environmental Protection report to

Figure 2–1.
Institutional Arrangements for Environmental Protection in Poland before January 1, 1999

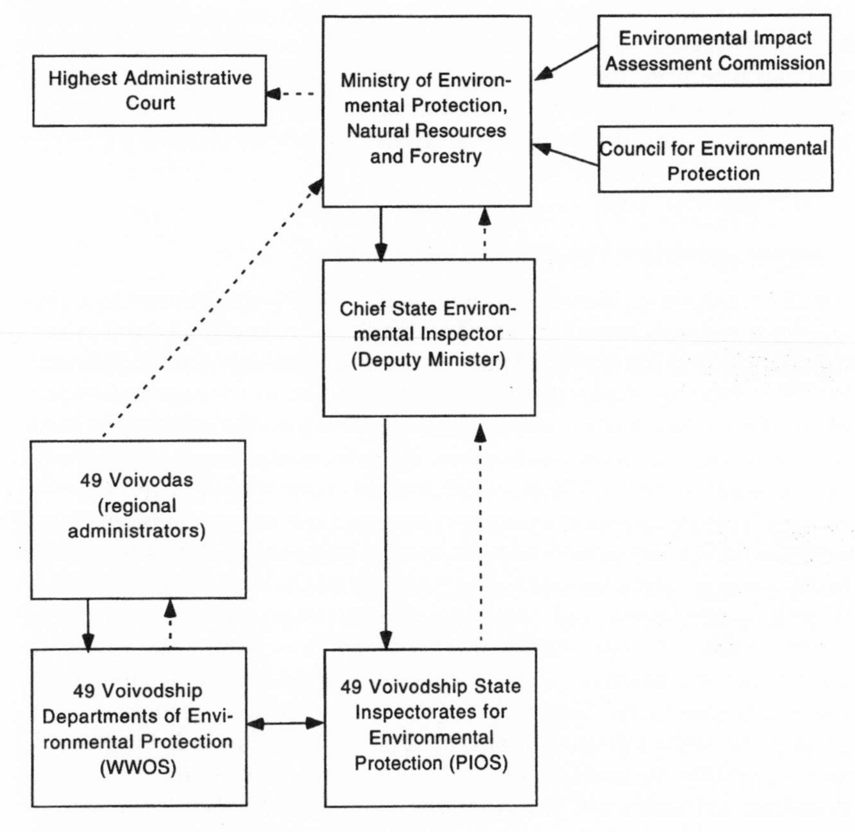

solid arrows: lines of authority
broken lines: appeals of administrative decisions

voivodas. The 49 enforcement offices are, in contrast to the licensing sector, independent of voivodas. They are integrated vertically into an independent organization, the State Inspectorate for Environmental Protection (PIOS), headed by the deputy minister of environmental protection.

The success of this administrative system depends critically upon the resources available to the permitting and enforcement authorities, the level of technical expertise of regulators, the level of access to and familiarity with industrial enterprises under their jurisdiction, the level of coordination among local, regional and national authorities and so forth. We report on these issues in later sections of this book.

A Standardized System of Licensing and Permitting for Industrial Enterprises

Poland maintains a set of detailed administrative procedures for permitting industrial facilities (see, for example, Jendrośka 1993; Environmental Protection Agency 1995; Cummings 1994). These procedures apply to both domestic and foreign firms.

Introduced during the 1970s and 1980s, the procedures have evolved since the passage of the 1980 Environmental Protection and Development Act through a series of administrative orders by the minister of environment protection. Since 1989, the licensing procedures have been codified and made more transparent. The primary regulatory tools are mandatory use and discharge permits (without which plants are forbidden to operate), ambient standards and environmental fees. Facilities are required to obtain from regional environmental administrators permits for water use, discharge of wastewater or sewage into ground or surface water, air emissions and disposal of solid waste. Permits are issued for a term determined by the administrator, commonly three years.

Prior to 1991, the information required in a permit application was at the discretion of the administrator and varied greatly among regions and across factories. Since 1991, applications for permits for water use, air emissions and waste disposal have had a standard detailed form, including descriptions of the technology in question and the environmental impact of the activity (the so-called *Operat*). The content of the permit application is specified in detail by the ministry. Thus, water use applications must include a description of the technology, its use and purpose, a blueprint of the factory, water mass balance analysis and, for groundwater discharge, hydrogeological analysis. For water discharges the application must also identify individual pollutants and show how each of 57 ambient water quality standards, will be met. By way of comparison, the United States EPA's so-called Ambient Water Quality Criteria are not legally enforceable water quality guidelines that could be used to grant water discharge permits.

Air permit applications must contain a description of the technology in question, identification of each source of emissions and the pollutants associated with that source, including individual emissions rates, description of pollution control technology and the predicted impact on air quality in the area. The analysis must also show if any of the 44 ambient air quality standards (30-minute peak concentrations, 24-hour and annual averages) will be exceeded and, in the case of predicted violations of standards, propose a detailed technical plan for reducing the emissions.

The relatively large number of legally enforceable air quality standards in Poland makes the permitting process highly standardized and relatively inflexible. Table 2–1, which compares Polish and U.S. standards for so-called criteria pollutants (a term taken from the U.S. Clean Air Act), of which there are only six, also shows that despite differences in averaging times, which makes comparisons somewhat uncertain, Polish standards are significantly more stringent, especially in protected areas such as national nature preserves. Polish standards for 31 toxic pollutants are generally less stringent than their counterparts in Massachusetts, one of the leading states in the United States to develop such guidelines and known for their strictness (Table 2–2). However, since in contrast to Poland the Massachusetts numbers are *non-enforceable* guidelines, these differences must be interpreted with caution.

The codification of water and air permit applications since 1991 has led to the emergence of a consulting industry specializing in this activity and to a more general professionalization of environmental impact analysis. The ministry maintains a list of licensed consultants, and although factories are not required to draw on that list, it is often to their advantage to do so. As a result, the licensed *Operat* specialists (of whom there were less than 100 nationwide as of 1998) have assumed the role of mediators between factories and regional authorities: they enjoy a privileged position with the licensing agencies and offer factories their experience and track record. Our case studies shed light on how this works in practice.

Applications for solid waste disposal are as detailed as those for water and air discharges, including technical specifications of the disposal site, a description of the volume and type of waste, the method of site management and a description of environmental impacts.

Permits for air and water discharges specify emission rates for individual pollutants from individual point sources. The procedure for calculating air emission rates starts with subtracting the background rates of a given pollutant in the given area from the allowable ambient air standards issued by the minister (in 1990 there were 45 such air standards). Regional public health authorities, Sanitary Inspectorates (whose functions are described in Chapter 3), determine the background levels either through direct monitoring or by using default values (based on a fraction of an occupational standard for a given pollutant). The so-calculated incremental allowable level of a given pollutant is converted back to an emission rate using dispersion models, accounting for local conditions and population centers. In highly polluted areas this translates into a very strict licensing regime, especially for criteria pollutants which start out with rather strict ambient limits.

Not only is the risk-based permitting system rather strict, it is also—at least on paper—very inflexible. Because the ambient standards are issued by the Ministry of Environmental Protection and the background levels are determined by the regional sanitary inspectors who report to

Table 2–1.
Comparison of Polish Ambient Air Standards with U.S. Primary National Ambient Air Quality Standards (for criteria pollutants)

	Poland						US NAAQS	
	All areas (ug/m3)			Specially protected areas (ug/m3)				
	30 min average	24 hr average	Annual average	30 min average	24 hr average	Annual average		
Carbon Monoxide	5000.0	1000.0	120.0	3000.0	500.0	61.0	8 hour average 1 hour average	10 mg/m3 40 mg/m3
Nitrogen Dioxide	500.0	150.0	50.0	150.0	50.0	30.0	Annual arithmetic mean	100 ug/m3
Ozone	100.0	30.0	-	50.0	20.0	-	8 hour average	0.16 mg/m3
Lead	3.5	1.0	0.2	2.0	0.5	0.1	Quarterly average	1.5 ug/m3
Particulate Matter (PM-10)	-	-	-	-	-	-	24 hour average Annual arithmetic mean	65 ug/m3 50 ug/m3
Sulfur Dioxide	Pre-1998: 600.0 Post-1998: 440.0	Pre:1998: 200.0 Post-1998: 150.0	Pre-1998: 32.0 Post-1998: 32.0	Pre-1998: 250.0 Post-1998: 150.0	Pre-1998: 75.0 Post-1998: 75.0	Pre-1998: 11.0 Post-1998: 11.0	3 hour average 24 hour average Annual arithmetic mean	80 ug/m3 365 ug/m3 80 ug/m3

the Ministry of Health, the regional environmental administrators (WWOS) have very little de jure flexibility in negotiating the terms of the air permits. In practice, their only options are (1) to delay a decision on issuing a permit while working with a facility on solving the emissions problem (regulators may issue temporary permits to enterprises that submit a plan for bringing emissions into compliance with permitted levels) or (2) to defer the problem to the enforcement branch (PIOS), which under the current regulatory regime has some flexibility in negotiating the compliance schedule (though not its terms). Regulators may defer fines for three to five years if a firm commits to rectifying violations (and will subsequently waive the fines if requisites changes are made).

With respect to water discharge permits, the licensing authorities have somewhat more flexibility. The regulations allow factories to exceed water quality parameters if it can be shown that the application of best available technology and redesign of the process would be inadequate to solve the problem and if it is determined that the public interest is best served by allowing the particular activity to continue. Permits for wastewater and sewage discharge into state or municipally owned treatment facilities are issued by the facilities themselves and the terms are negotiated between the factories and the facilities directly. Strict adherence to water quality standards is not required. Here too the regulations are specific as to the types and concentration levels of individual pollutants, locations of monitoring stations, fees and fines. Permits for hazardous waste are issued by WWOS in consultation with the regional sanitary inspectors. The terms of the permits are negotiated on a case-by-case basis.

Siting of new facilities follows a somewhat different regulatory pathway, one that attempts to draw on local, regional and central authorities and to open the process to public participation. According to the Land Use Planning Act of 1984, any new development requires planning permission. The procedure has two stages: obtaining preliminary consent ("site indication") for the proposed type of development and obtaining a final permit ("location decision"). The preliminary consent is issued by voivodship and local authorities, except in cases involving certain major projects, such as large power stations or facilities that are potentially especially damaging to the environment and public health. In practice, the first stage of the siting process begins with an investor approaching the local government (*gmina*) with an application to site the industry in the community. The *gmina* reviews the application in the context of the community plan for socioeconomic development, zoning, types of economic activity already present and the type of activity the community values. If such a plan does not fit within the community plan, the plan can be amended. The preliminary consent is a legislative act on the part of *gmina*, requiring approval of the board of selectmen.

Table 2–2.
Comparison of Polish Ambient Air Standards and Massachusetts Ambient Air Guidelines

	Polish Ambient Air		
	All areas		
	30 min average	24 hr average	Annual average
Acrylonitrile	7	2.0	0.5
Acetyl aldehyde	20.0	10.0	2.5
Methanol	1000.0	500.0	130.0
Ammonia	400.0	200.0	51.0
Aniline	50.0	30.0	10.0
Arsenic	0.2	0.05	0.01
Benzene	35.0	10.0	2.5
Benzoapyrene	20.0 ng/m3	5.0 ng/m3	1.0 ng/m3
Chlorine	100.0	30.0	4.3
Vinyl chloride	16.0	5.0	1.3
Chromium	8.0	2.0	0.4
Tetrachloroethylene	600.0	300.0	70.0
Dichloromethane	400.0	150.0	60.0
1,2-dichloroethane	400.0	150.0	60.0
Carbon disulfide	50.0	20.0	3.8
Phenol	20.0	10.0	2.5
Fluorine	30.0	10.0	1.6
Formaldehyde	50.0	20.0	3.8
Cadmium	0.22	0.22	0.01
Xylene	300.0	100.0	16.0
Sulfuric acid	200.0	100.0	16.0
Hydrogen chloride	200.0	100.0	20.0
Copper	20.0	5.0	0.6
Nickel	340.0 ng/m3	100.0 ng/m3	25.0 ng/m3
Nitrobenzene	50.0	30.0	10.0
Mercury	2.0	0.3	0.04
Hydrogen sulfide	30.0	5.0	1.0
Styrene	20.0	7.0	2.0
Toluene	300.0	200.0	50.0
Vanadium	3.5	1.0	0.25

Standards (ug/m3)			Massachusetts Ambient Air Limits	
Especially protected areas				
30 min average	24 hr average	Annual average	24 hr average	Annual average
7	2.0	0.5	0.4	0.01
10.0	5.0	1.3	2	0.5
200.0	100.0	25.0	7.13	7.13
100.0	50.0	13.0	100	100
20.0	10.0	2.5	0.2	0.1
0.2	0.05	0.01	0.0005	0.0002
35.0	10.0	2.5	1.74	0.12
20.0 ng/m3	5.0 ng/m3	1.0 ng/m3		
30.0	10.0	1.6	3.95	3.95
16.0	3.0	0.4	3.47	0.38
2.5	0.5	0.08	(metal) 1.36 (compounds) 0.003	(metal) 0.68 (compounds) 0.0001
200.0	120.0	30.0	922.18	0.02
100.0	60.0	15.0	9.45	0.24
100.0	60.0	15.0	11.01	0.04
15.0	4.5	0.6	0.1	0.1
10.0	3.0	0.4	52.33	52.33
10.0	3.0	0.4	- (fluoride 6.80)	- (fluoride 6.80)
20.0	10.0	2.5	0.33	0.08
0.2	0.2	0.001	0.003	0.001
40.0	10.0	1.3	(xylenes) 11.80	(xylenes) 11.80
100.0	50.0	7.9	2.72	2.72
100.0	50.0	10.0	7	7
6.0	2.0	0.3	0.54	0.54
340.0 ng/m3	100.0 ng/m3	25.0 ng/m3	(metal) 0.27	(metal) 0.18
20.0	10.0	2.5	13.69	6.84
0.4	0.1	0.02	(elemental 0.14 (inorganic) 0.14	(elemental 0.07 (inorganic) 0.01
4.0	1.0	0.5	0.9	0.9
10.0	3.5	1.0	200	2
100.0	50.0	13.0	80	20
0.1	0.1	0.0005	0.27	0.27

The application for an operating permit is reviewed only by the local and voivodship authorities. An application must be accompanied by detailed information about the project and its environmental impacts, including a response to the conditions in the preliminary consent. For "especially harmful" projects an Environmental Impact Statement is required. The *voivoda* can also, at its discretion, require an EIS for projects not classified as "especially harmful." This is one of the few areas where a *voivoda* has formal discretionary authority. At this stage the voivodship Department of Environmental Protection prepares individual environmental permits and determines the pollution fees while the *gmina* considers the final permit. Once construction is complete, the investor must secure the usual permits for air, water and waste from appropriate authorities: from WWOS, which (in principle) should be a simple matter since WWOS has been involved in the process throughout; from local and regional public health authorities; and from the fire marshall. These authorities perform facility inspections to ensure that the initial conditions of the construction permit were satisfied. Fourteen days after the site visit, the authorities must notify the investor of any problems. If no response is made, the facility gains automatic approval. If the voivodship Department of Environmental Protection approves an application on environmental grounds the local government cannot overrule the decision; but if the application is denied by WWOS, the *gmina* may challenge the decision before the Administrative Court of Appeals.

Overall, the approach to permitting both existing and new factories is highly regulated and rigid towards both the licensing authorities and the firms, allowing relatively little discretion for negotiating conditions of permits. At its heart is a traditional pollutant-specific/medium-specific/standard-based permit. The strict reliance on ambient air quality and water standards issued by the Ministry of Environmental Protection, the relative strictness of many standards and the linking of permissible discharge rates to the existing background levels of individual pollutants often places stringent demands on firms. For this reason, many researchers continue to voice concern as to the effectiveness of the permitting system (see, for example, Sleszyński 1998). Adamson et al. (1996), Żylicz (1994) Toman (1993) and Toman, Cofaba and Bates (1994), among others, have suggested that a greater use of economic instruments (such as energy taxes, tradable pollution schemes) would improve the economic efficiency of the system (i.e., achieve the same environmental results at lower cost). Further, it is unclear how many firms are operating without valid permits. Sleszyński (1998) and Warner (1996) cite an estimate that 50% of registered firms are operating without valid permits, although there are little reliable data on which such statistics can be based. Our own research suggests that the proportion of firms operating without permits is actually far lower.

Extensive Monitoring and Reporting Requirements

In addition to obtaining discharge licenses, facilities are required to comply with extensive monitoring and reporting requirements. A factory using groundwater must report quarterly water use levels and any changes in the state of the aquifer or in technology employed. Discharges of wastewater must be reported annually to the National Bureau of Statistics and the voivodship environmental office (WWOS). These reports include water volumes, concentrations of contaminants and location of discharges. The reporting requirements for air emissions and solid emissions and solid waste disposal are similarly extensive, including lists of substances, individual sources within a facility and rates of generation and emission. The regional WWOS offices and the ministry maintain these data, both facility-specific and aggregate for voivodships and countrywide.

Emissions and waste discharges are based on a system of self-reporting by the firms involved. This of course raises the question of whether firms have the technical expertise to meet these reporting requirements and whether they provide accurate information to regulators. On the other hand, the system for monitoring ambient air quality and surface and groundwater conditions has been substantially strengthened.

A Well-Established System of Pollution Charges and Fees

Pollution and resource use (fees) in Poland date to 1974, well before almost all other industrialized countries (Jendrośka 1996b). Fees, set by the Ministry of Environmental Protection, are applied to a wide range of activities (including air emissions, waste storage, water diversion and consumption, waste discharges and use of agricultural land for non-agricultural purposes) and are collected by WWOS. The fees are based on volumes (waste disposed, discharge and emission rates of individual pollutants, water use) and are calculated on the basis of the information provided in permit applications. As discussed earlier, the revenues from the fees are channeled into the national, voivodship and local Funds for Environmental Protection and Water Use.

Environmental fees have been increasing over time, as illustrated in Table 2–3 for selected air pollutants and in some cases, such as sulfur dioxide, are among the highest in the world (Jendrośka 1998, p. 93). Figure 2–2 indicates that national revenues from the fees have grown significantly between 1991 and 1997. There can be little doubt that environmental fees are an important source of revenue to be channeled into investments in environmental protection through the Environmental Fund. For example, revenues from fees on SO_2 emissions in 1996 were approximately U.S.$126 million (Sleszyński 1998).

Table 2–3.
Changes in Emission Fees for Selected Air Pollutants (złoty/kg* and approximate U.S.$ values)**

Effective Date	Lead	Sulfur dioxide	Benzene
Jan 1, 1991	36,000 ($38)	680 ($0.07)	1800 ($0.21)
Jan 1, 1992	500,000 ($38.50)	1,100 ($0.08)	1,000,000 ($76.90)
Jan 1, 1993	500,000 ($31.20)	1,100 ($0.07)	1,000,000 ($62.50)
Jan 1, 1996	79.04 ($36)	0.25 ($0.10)	158.08 ($72)

*Prices for 1991–1993 are denominated in old złoty; for 1996 in new złoty. One new złoty is equivalent to 10,000 old złoty.
**From Cole (1997) and from the Environmental Protection Agency (1996).

For some firms the scale of environmental fees can be quite high. A survey of 112 of the largest polluters in Poland found that pollution charges constituted 4.9% of total corporate expenditures (Sleszyński 1998). Nevertheless, the role of fees as an economic incentive in directly promoting changes in technology and industrial activity is unclear. In an analysis of SO_2, Anderson and Fiedor (1997) concluded that the marginal cost of meeting 1998 emissions standards for large enterprises was still far above the current fee level. Other authors also concluded that for most enterprises the fees for pollution are below the level necessary to create significant economic incentives for pollution prevention (Sleszyński 1998; Bluffstone and Larson 1997; Toman et al. 1994).

Environmental fees that are imposed are generally paid, but there are important exceptions. For example, the mining industry continues to refuse to pay environmental fees or to meet required environmental standards. This industry is illustrative of the segments of heavy and primary industry where substantial economic and environmental restructuring is required but where the social costs of restructuring, especially at the regional level, are very large. Outside of this sector, industry has made no broad challenge to the legitimacy of these environmental policy instruments.

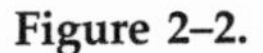

Figure 2–2.
Environmental Fees Collected in Poland between 1991 and 1997 (in millions of złoty*)

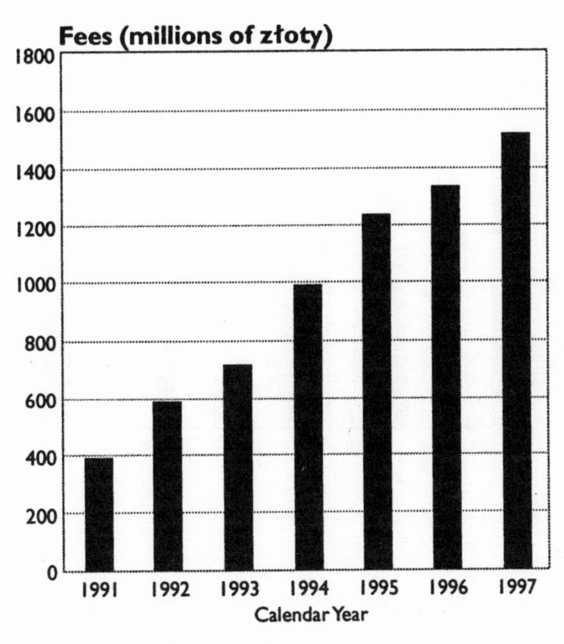

*Amounts expressed in new Polish złoty.
Source: GUS 1998b, p. 465.

The percentage of fees that are collected has declined as the level of fees has increased, but for most types of fees, the collection ratio remains reasonably high. Figure 2–3 shows that the collection success was reasonably good between 1991 and 1997, ranging between 62% and 75% for all fees combined and between 80% and 94% for air and water pollution fees. Since these statistics include the recalcitrant coal mining sector, it appears that the rest of industry has a good compliance record regarding the payment of pollution charges.

A Strengthened Regime of Enforcement and Pollution Fines

The enforcement system has been considerably strengthened since 1989. It is now carried out by a politically strong organization which commands substantial resources, broad powers and effective regulatory tools. First created in 1980, the State Environmental Protection Inspectorate (PIOS) has grown in stature. In 1991 the position of the chief environmental inspector was elevated to minister of environmental protection. Enforcement activities are carried out by the chief inspector

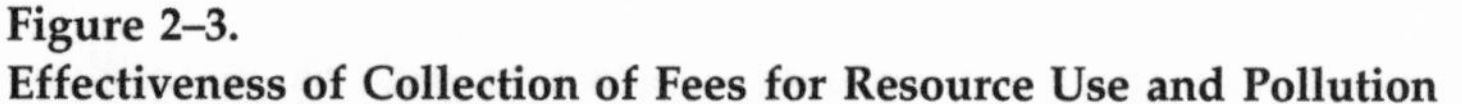

Figure 2–3.
Effectiveness of Collection of Fees for Resource Use and Pollution

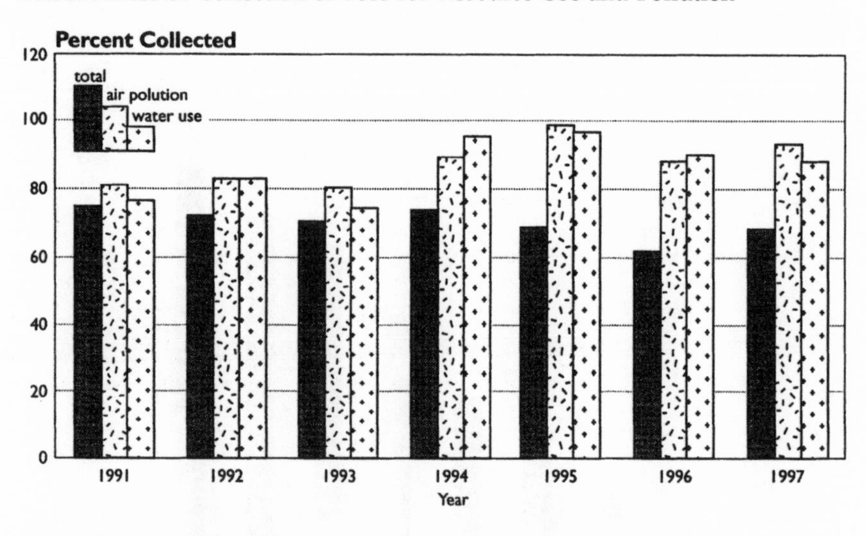

Source: GUS 1998b, pp. 465–470.

through 49 voivodship PIOS offices, which report back to the inspector (rather than to the regional administrator, as was the case before 1991). The chief environmental inspector hears appeals of decisions of voivodship inspectors while the minister hears appeals of the chief environmental inspector's decisions (Figure 2–1).

The voivodship inspectors have the power to impose non-compliance fines, which have been substantially increased since 1992; to halt activities that endanger the environment, including closing of factories; to ban sales and import of raw materials and other products that fail to meet Polish environmental standards; and to recover inspection costs if non-compliance is discovered through such an inspection. In addition, the inspectorate is responsible for running a nationwide environmental monitoring network.

Since 1990, PIOS has intensified its activities. The number of firms it covers has grown from approximately 31,000 to 37,000 between 1993 and 1997, and the number of inspections increased by 5% during that period. The number of orders to make improvements and number of fines imposed have also grown between 1993 and 1997: by 24% and 32%, respectively (Stodulski 1999, pp. 52–53, based on the PIOS statistics).

Pollution fines are one of the key instruments of enforcement in Poland. Unlike pollution fees, which are budgeted into operating costs, fines are paid from after-tax income. The revenues from non-compliance fines are deposited in the same national funds for environmental protec-

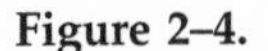

Figure 2–4.
Environmental Fines Collected in Poland between 1991 and 1997 (in millions of złoty*)

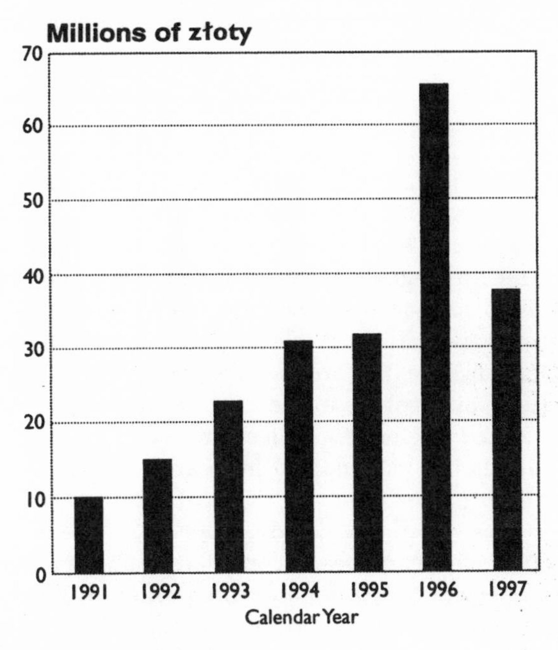

*Amounts expressed in new Polish złoty.
Source: GUS 1998b, p. 465.

tion as the fees. Historically, Poland was among the first industrialized countries to implement economic incentives for pollution prevention by instituting (in the 1970s) fines for non-compliance (Jendrośka 1996b). During the communist period these fines were symbolic in magnitude and often ignored by all parties involved in enforcement.

Since 1989 the ministry has repeatedly increased the fines, and Figure 2–4 suggests that they are an important source of revenue for the National Environmental Fund. According to the National Bureau of Statistics (GUS 1998a), the collection rate for fines has been very low, around 13%. For two reasons, however, these statistics provide little useful information about the behavior of "average" privatized firms, such as those chosen for our study. First, nationally, most of the fines are imposed on the mining sector and on the large state-owned "dinosaur" enterprises that are de facto bankrupt and cannot pay the fines. Second, payment of fines can be legally deferred (or forgiven altogether, under some conditions) in favor of investments into pollution prevention (see below). Such deferred or waived fines appear in GUS statistics as "uncollected." Nonetheless, there is a well-founded concern that a shortage of inspectors and monitoring tools, and the uneven technical expertise of small and medium-size firms, may interfere with effective enforcement (Weber 1996; Millard 1998). Our case studies (Chapters 4 and 5) and survey (Chapter 6) shed more light on the questions of compliance with standards and collection of fees and fines.

Discretion and Flexibility

In the current industrial facility permitting system, the voivodship licensing authorities have limited explicit discretion to balance environmental goals against other competing priorities, such as maintaining employment. The only legally sanctioned regulatory relief mechanism at their disposal is to delay a licensing decision following the submission of a formal permit application. While the law does not allow facilities to operate without a license, the process does implicitly recognize the use of this strategy by authorities. In the case of air pollution a temporary permit can be issued; the emissions fees are then doubled, but since no licensing conditions are violated, no fines are imposed. In effect, fines are traded for fees.

This fees-for-fines swap may be advantageous to the firm. First, doubled fees may be lower than fines. Second, fees count as part of operating costs, whereas fines are paid from after-tax profits. And yet the practice of delaying issuance of a license is clearly not an optimal solution. It undermines the legitimacy of the law and due process, and creates an incentive to submit incomplete permit applications. It also perpetuates the discredited past practices of disregard for formal policies. Not surprisingly, the representatives of the voivodship WWOS we interviewed were openly critical of their lack of discretion in determining the conditions of permits.

One of the more significant directions of post-1989 reforms addresses this issue of flexibility in negotiating operating licenses. Through the so-called Compliance Schedule Program, regulators will seek to replace (for certain industrial sectors) the ambient air standard approach with nationwide emission standards; and for factories that cannot meet the emission standards, regulators will be able to negotiate the terms of environmental permits (Bell and Bromm 1997). In essence the proposed reforms recognize the need to balance environmental protection with other social objectives, address the vulnerability of the current enforcement system and remove some of the drawbacks of the centralized command and control regulatory scheme. Implicitly, the reform would legitimate the informal procedures that have been quietly implemented on the regional level over the years.

However, the reach of these reforms is limited. The compliance schedules would apply only to certain sectors, and in practice they have been designed for major point-source polluters. In many cases the Ministry of Environment will likely retain direct control over these negotiations. Most small and medium-size firms will likely not qualify for the program. It is therefore likely that two systems of environmental management will emerge: one, a system of explicit negotiations for large and economically significant polluters and the other, a system of limited regulatory discretion and less formal balancing of objectives. This reluctance by the central authorities to share power with regional administration is, according to Jendrośka (1998), typical of Poland's bureaucratic tradition.

In contrast to the licensing branch, the enforcement authorities (PIOS) enjoy greater explicit discretion to provide relief to economically fragile factories that show good-faith efforts toward pollution abatement. They have the authority to grant up to five years of delay or forgiveness of non-compliance fines. The conditions for obtaining such an extension are the submission of an approved plan for pollution abatement and demonstration of financial inability to pay (as judged by the authorities). If the plan is successfully implemented by the approved deadline, the funds invested in environmental improvements are applied towards the non-compliance fees.

The voivodship PIOS representatives we met were pleased with their system of explicit negotiations and sounded confident about their ability to implement the system. "We have never had a firm default on us in implementing an improvement plan we have negotiated," remarked a director of a voivodship PIOS. Among the representatives we interviewed, on both the licensing and enforcement sides, a strong preference for negotiations over confrontation was consistently voiced. Typical remarks were: "A repressive system does not work," and "When we see good faith on the part of management, we try to prevent their financial collapse."

During our interviews with voivodship WWOS and PIOS authorities we also observed evidence of extensive cooperation between the licensing and the enforcement branches to provide relief to financially strapped firms that were trying to make environmental improvements. In one of the voivodships the representatives of the two agencies described their informal understanding regarding the apportioning of fines and fees through the terms of the air pollution license for such an enterprise: WWOS would increase the allowable emission rates, which would generate higher pollution fees but make it easier for a factory to achieve compliance (or at least, in the case of non-compliance, lead to lower penalties). While cooperation between PIOS and WWOS is not surprising, it is not legally required, and its apparent frequency suggests the existence of a more complex and informal system of regulating industrial enterprises than would be disclosed by a study of formal laws and policy documents. (We explore this issue in Chapters 4 and 5.)

Limited Public Participation or Organized Industrial Involvement in the System

The 1980 Environmental Policy and Development Act was quite progressive on the issue of public participation. It explicitly granted NGOs the right to file public interest lawsuits and to access information about firms. Since 1989 the legal and procedural record with regard to public participation has been mixed. During the early 1990s the public access to information about individual firms has been actually curtailed. The central bureau of statistics, which collects firm-specific data, releases only the aggregate data for industrial sectors. On the other hand, the provisions for public participation in licensing new facilities and in the Environmental Impact Assessment process indicate a potential for change. There is also a strong sentiment among some drafters of the amended Environmental Protection and Development Act (who, ironically, have debated since 1997 largely behind closed doors) to strengthen the public participation in policy making and implementation and to enhance broad input into policy development. For example, the 1999 discussion draft of the act provides for the right of individuals to demand administrative action against polluters. It also mandates the ministry to hold a three-week open comment period before finalizing any major rulings. In an unprecedented move, in November 1998 the Ministry of Environment actively sought comments on the draft text of the proposed legislation among the industry groups, local and regional environmental authorities and environmental organizations, following it in March 1999 with a forum to discuss the responses. Several dozen organizations and individuals participated.

In practice, however, there are still serious obstacles to broad participation in policy making and implementation, especially by the general public. First, the law challenges both the bureaucracy's deeply entrenched administrative resistance to external scrutiny and its disdain for the value of lay persons' contribution to data analysis and policy making. Second, the independent ecological organizations have no traditions of participative legal process and are too fragmented to mobilize their limited resources necessary for such participation. For example, of the 60 environmental NGOs invited to participate in the March forum only five did. Third, all parties are strongly influenced by the prevailing cultural mores, which, in Poland, favor delegating problems to experts who solve them in close meetings. One of the drafters of the 1997 Environmental Protection and Development Act, for example, made a revealing—if informal—comment during a discussion of the proposed provisions for public access to information: "If we must, let us put the provision in, and let us hope that the public does not use it."

Similarly, organization within industrial sectors has been slow, with the exception of large energy and extraction sectors, which have effectively participated in policy making and implementation since 1989. For the most part firms continue to be recipients of regulations rather than participants in their formulation. But this is changing. A surprising number of small and medium-size enterprises, especially paper and packaging manufacturers, participated in the ministry-sponsored forum on the proposed amendments to the Environmental Protection and Development Act, submitting extensive written comments. During a spring 1999 interview president of the Confederation of Polish Employers told us of a growing interest among firms to challenge administrative decisions in court, using services of specialized law firms and technical consultants.

As we discussed earlier, the national NGOs have an interest in participating in national policy making, although their record there is mixed. NGOs' interest in enforcement and performance issues on a firm level is, however, negligible. There is also little evidence of institutionalized participation by voivodship WWOS and PIOS organizations in developing national environmental policy, although both have nominal representation among the central authorities. The March 1999 forum was a significant break with that practice. Notably, the regional authorities were active participants in the process, which may indicate that changes are forthcoming.

CHANGES IN THE SYSTEM SINCE JANUARY 1999

On January 1, 1999, the governance system in Poland was decentralized. The 49 administrative regions of the central government (voivodships) were consolidated into 16 voivodships whose power has been reduced, while many local decisions have been delegated to the 373 newly created (or, rather, recreated from the pre-1975 period) elected county governments (*poviat*) led by *starosta*. These changes also affect the institutional arrangements of environmental (and occupational) protection.

Figure 2–5 depicts the new institutional arrangements and lines of authority, although the question marks indicate that not all the arrangements are clear at the time of this writing. One of the major changes involves the shift of the facility licensing functions from the Voivodship Departments of Environmental Protection (WWOS) to the county level (*poviat*). While the offices of *voivodas* will most likely maintain divisions of environmental protection, these will be advisory bodies. In short, WWOS has been eliminated. The authority to review and approve new and existing facilities now rests with 373 county (*poviat*) departments of environmental protection which report to the elected county governing bodies, each headed by a *starosta*. As indicated in Figure 2–5, the direction of the appeals of the licensing decision—the *voivoda* or the minister—is not clear at the time of this writing.

Figure 2–5.
Institutional Arrangements for Environmental Protection in Poland after January 1, 1999

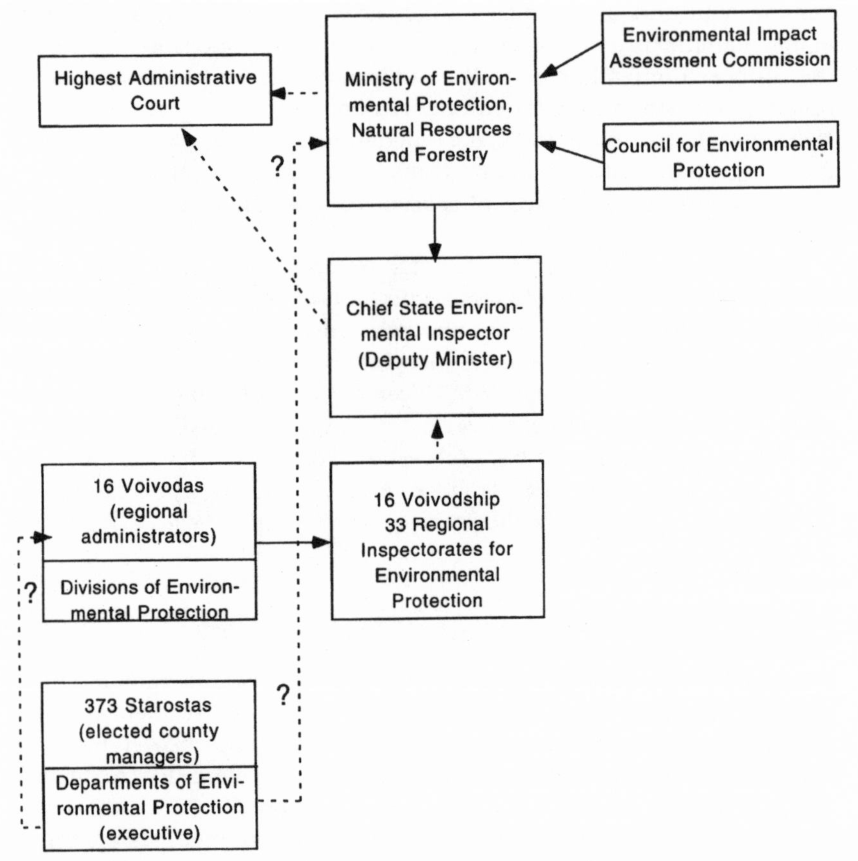

Another major change concerns the enforcement branch. The 49 voivodship PIOS offices have now been replaced with 16 voivodship and 33 regional offices. The recently acquired independence of the regional inspectors, in existence only since 1991, is now partially curtailed back to the pre-1991 status: instead of reporting directly to the chief environmental inspector, they have become part of the *voivoda's* office. On the other hand, the appeals of the regional inspectors' decisions are, as in the recent past, heard by the chief environmental inspector. The chief inspector's decision can be in turn appealed to the highest administrative court, not the minister of environmental protection (see Figure 2–1).

The shift of the facility licensing authority away from the central government (through the regional offices) to the local elected governance, and from the regional level to the county level, is consistent with the current practice of case-specific flexible decision making that is grounded in sustained interactions between the authorities and the firms. On the other hand, the need to staff the large number of new environmental departments with technical personnel will increase the fiscal pressures on the local level (part but not all the necessary funding will continue to be allocated from the state budget).

The curtailment of the regional inspectors' independence is of potential concern. However, during the interviews in the spring of 1999 regional environmental directors told us that the changes are unlikely to affect the work of the environmental inspectors, partly because the appeal process continues to be independent and partly because the changes are consistent with of the well-established collaborative relationship between the licensing and enforcement authorities. Only time and empirical evidence will show the true impact of these changes on the implementation process.

PROSPECTS FOR EFFECTIVENESS

Throughout much of Central and Eastern Europe, the past decade has brought a substantial reduction in industrial pollution (Bluffstone and Larson 1997; Schnoor, Galloway and Moldan 1997). Many of these gains derive from reductions in output and economic dislocation and from industrial restructuring accompanying the transition from communist rule. But it is also likely that changes in systems of environmental protection have contributed to improvements in environmental quality.

In Poland, where by the mid-1990s GDP had already surpassed pre-transition levels, the quality of the environment has demonstrably improved. Figure 2–6 shows a drop in several physical indicators of pollution between 1990 and 1997. The dramatic reduction of industrial discharges into the rivers led to parallel improvements in water quality. For example, the proportions of Class I and II rivers considered clean have increased approximately two-fold during that period (Stodulski 1999, p. 40). As shown in Figure 2–7, the pollution intensity of economic activity has increased several-fold in Poland. These are the payoffs from technological investments in the energy sector, which have been systematically promoted by the government during the 1990s.

One problem with interpreting aggregate statistics is that changes in different economic sectors may counteract each other. For example, the growth of emissions from the transportation sector in Poland, shown in Figure 2–8, may camouflage larger reductions in emissions from point sources. Another crucial shortcoming of the aggregate statistics is that they do not identify the effects of regulatory reform on the behavior of

Figure 2–6.
Trends in Environmental Quality in Poland Relative to a Base Year

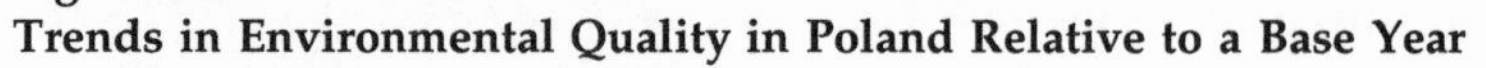
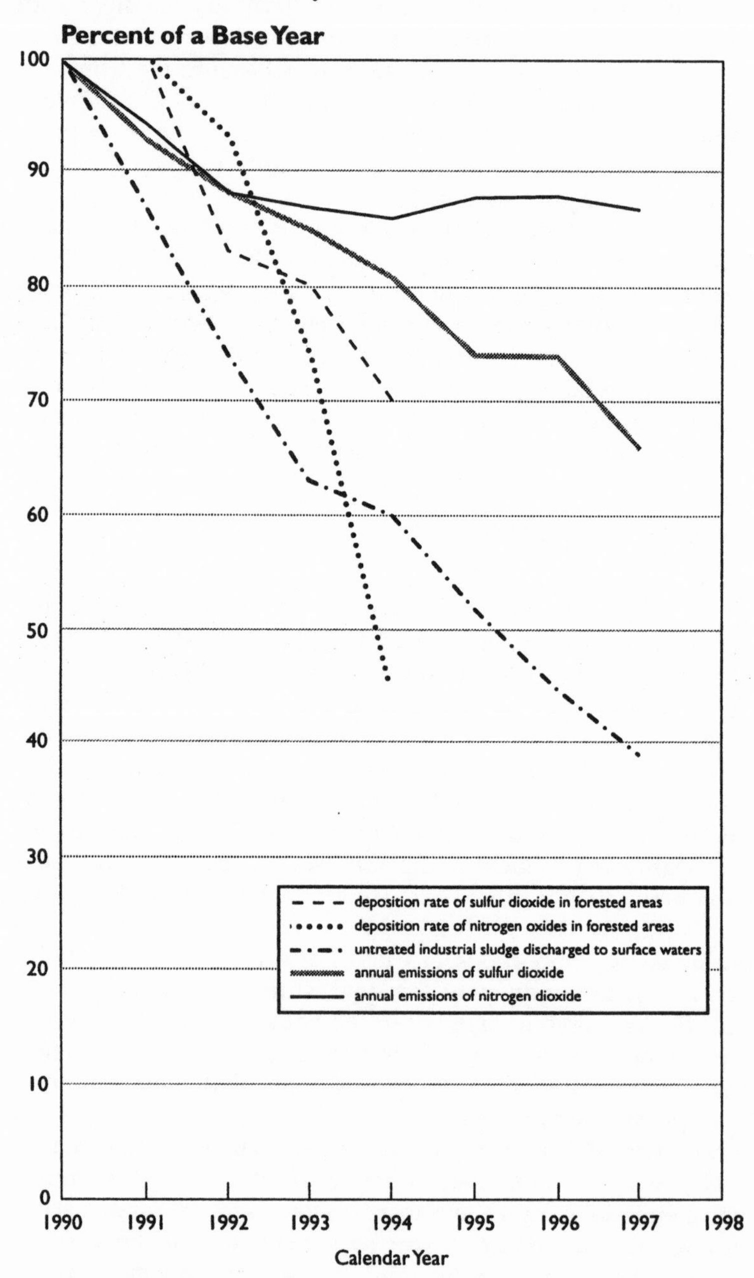

Sources: GUS 1996, pp. 112, 128, 260; GUS 1998a, p. 26; GUS 1998b, pp. 138, 307.

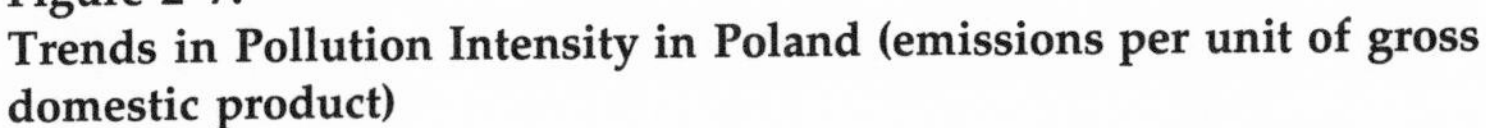

Figure 2–7.
Trends in Pollution Intensity in Poland (emissions per unit of gross domestic product)

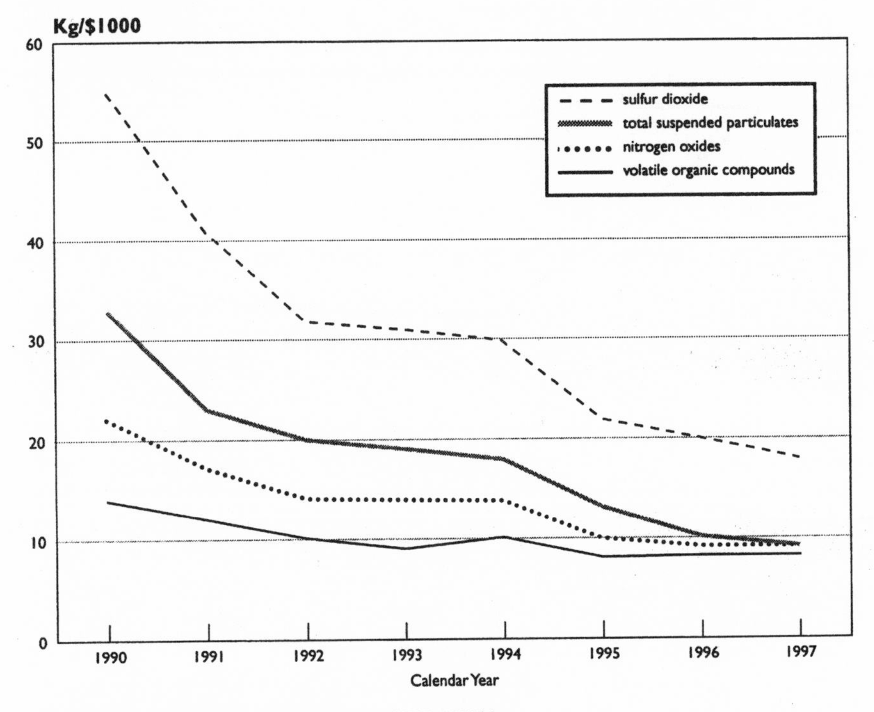

Source: Stodulski 1999, p. 35 (based on GUS 1998b).

individual firms, especially in the small and medium-size manufacturing range. Aggregate improvements in environmental quality may be due to the economic collapse of dirty industries, or to the mitigation of some of the worst point-source polluters, or to the ongoing transition from a manufacturing-oriented economy toward a service- and trade-oriented economy. The deceleration of the improvement rate in air emissions since the mid-1990s, shown in Figures 2–7, 2–8 and 2–9, is consistent with the first two explanations. Almost certainly, each of these three factors played a role in Poland. Growth and new capital investment have certainly reduced the significance of older "dirty" industrial stock within the Polish economy. Thus the World Bank estimates that at current growth rates less than 25% of Poland industrial capital stock in 2005 will be of pre-1990 vintage. Such rapid capital turnover can result in improved environmental quality without shedding much light on the effectiveness of the regulatory system in improving the environmental performance of individual firms.

Figure 2–8.

Trends in Emission Rates from Transportation in Poland

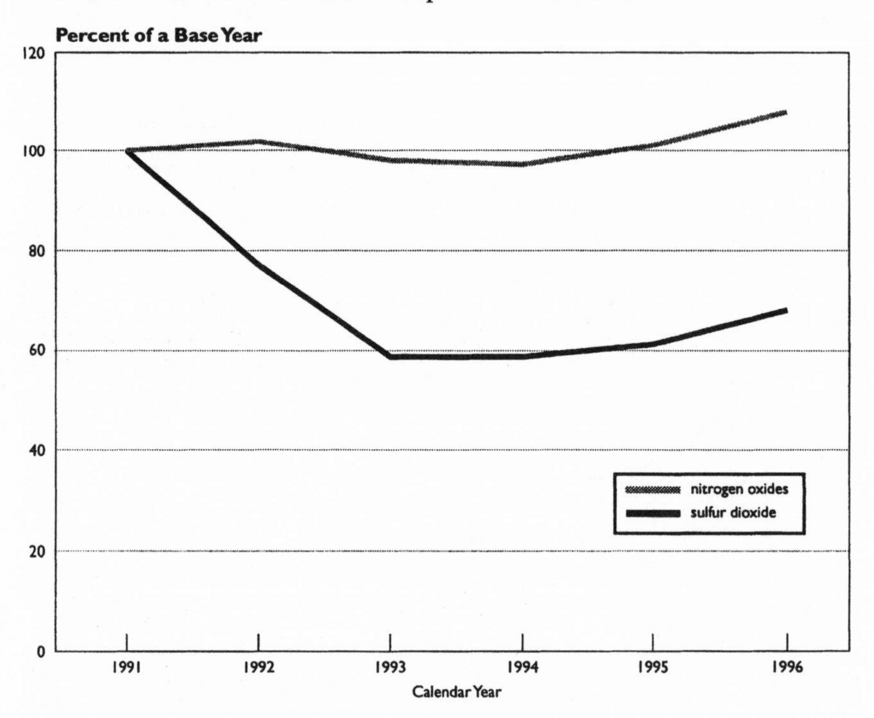

Source: GUS 1998b, p. 212.

Cole and Clark (1998) suggest that environmental protection in Poland is now moving from the "easy" stage, in which improvements are virtually automatic, to an "interesting" stage, in which the efficacy and efficiency of the regulatory system will be genuinely tested. Certainly, Poland entered the 1990s with a considerable foundation for environmental protection. Rather than creating a whole new system of environmental protection, Poland was able to draw upon an existing system of laws, policies and standards. In this context the main challenge was one of operationalizing the system, that is to say, bringing it to bear, in practical terms, on the activities of all industrial enterprises, from large polluters to small and medium-sized firms. Recent incremental reforms have clearly strengthened the laws, policies and institutional arrangements of the existing system, resulting in an approach to environmental protection that has a range of advantages in the current economic and political context.

The inherited system provides a comprehensive set of standards, enforcement practices, and permitting procedures, supported by an existing

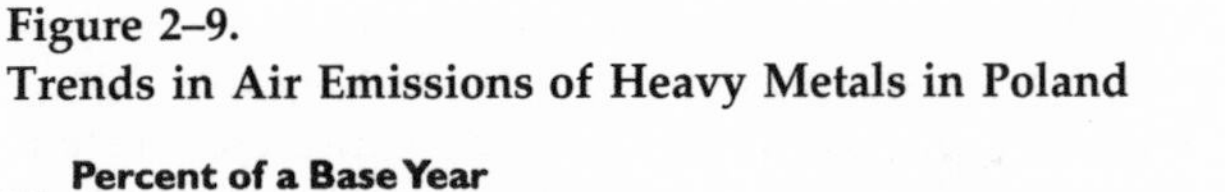

Figure 2–9.
Trends in Air Emissions of Heavy Metals in Poland

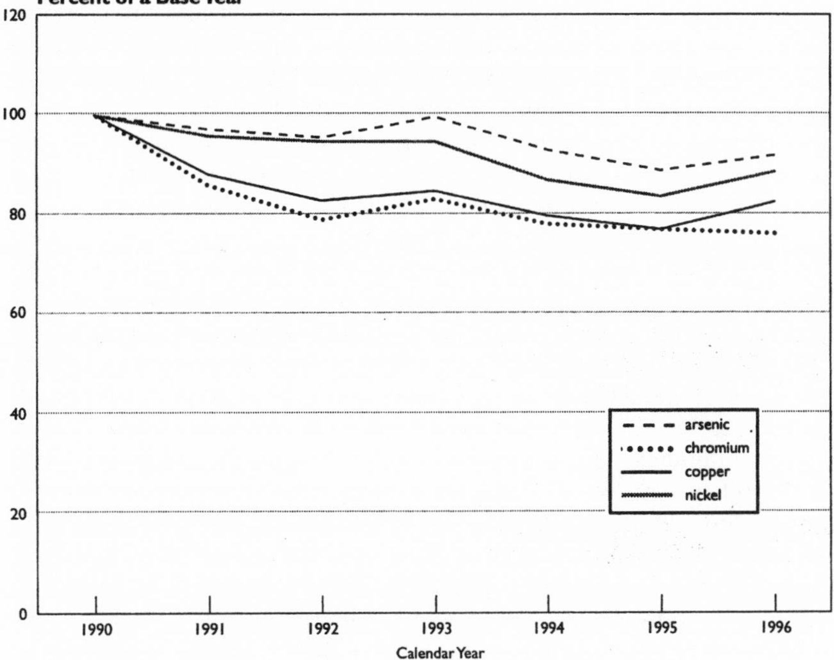

Source: GUS 1998b, p. 208.

institutional infrastructure and legal framework, and an increasingly decentralized system of monitoring and enforcement. Ambient standards, fees, fines, and the newly codified licensing process provide what appear to be transparent permitting procedures based on detailed technical analysis. While the system depends heavily on coercive policy instruments, such as licenses and fines for non-compliance, economic instruments such as fees, negotiated environmental permits and enforcement options and pollution trading schemes are growing in significance. Importantly, the system enjoys strong support from the academic and other technical elites and from the bureaucracy. The vitality and commitment of these groups is a continuation of pre-1989 practices. In fact from the early 1980s to the late 1990s, there has been a notable continuity among the individuals who play key roles in environmental policy making and innovation.

Regulators and policy makers have demonstrated a considerable capability to learn and to experiment with alternative policy approaches. Since 1989 there has been a proliferation of ideas regarding regulatory reforms and effective policy instruments. Some ideas, such as radical system restructuring, have been abandoned; others, such as negotiated enforcement and pollution trading systems, are being cautiously tested; and some, such as the National Environmental Fund and international debt-for-environment swaps, have been very successful. Several European approaches, such as generally applicable permits and negotiated license agreements, are being piloted or actively debated. In general, there appears to be sensitivity among policy leaders to choosing those policy instruments with greater promise of effectiveness. There are even deliberate attempts to provide more explicit guidance for balancing competing social objectives. This is evident in the language of the amended EPDA, in PIOS's increased discretion to enter into negotiated enforcement agreements with firms, and in the limited approaches to negotiated licensing of large enterprises. Evidence is also accumulating that the reforms have strengthened the enforcement of environmental regulations.

At the same time the environmental protection system in Poland places a heavy burden on enterprises to perform technical assessments, to comply with standards (many of which are very stringent), to monitor and report and to pay fees. Overall, the system remains highly resource intensive, depending upon extensive monitoring and assessment by firms and regulators alike of a large number of emissions standards. Despite recent reforms, the system continues to be rather inflexible, especially with respect to medium-sized and small enterprises. In fact the selective deployment of negotiated licensing creates incentives for the emergence of two systems: one for the largest firms, based on transparent negotiations, and the other, where balancing and negotiation are more difficult and less formal. The January 1999 decentralization of the facility licensing is likely to further contribute to this process. The strong support from the administration and intellectual circles notwithstanding, the environmental protection system has a relatively narrow support base. The limited participation in the recent reforms by most industrial sectors, by regional offices, by the grassroots ecological movement and by the public at large makes for an elitist system. These actors may be, or may become, participants in the system, but there is little reason to regard them (at least yet) as stakeholders. It is not clear how such a system would respond to an organized challenge.

The efforts to bring all firms formally under the system of environmental regulation faces obvious difficulties: the high demands on firms, limited administrative resources for licensing and enforcement, limited flexibility in balancing competing objectives in facility-specific decisions, economic pressures and lack of organized support network among the public. It seems reasonable to expect that some firms will resist the regulations. The legacy of the past disregard for official policies and standards may very well be lingering among the regulatory and enforcement institutions as well. This is an important point to consider. The entire reform program has, in effect, been premised on the plausible but untested assumption that when confronted with both systemic change in the political and economic domains and incremental reforms in environmental law and enforcement institutions, the key actors in the environmental protection drama—regulatory authorities and industrial managers alike—would in fact change their behavior.

It is because of this underlying assumption that the reform process focused principally upon the legal and institutional domains. The architects of reform targeted their efforts on creating modern institutions, policies and policy instruments. Less attention was given, either by the reformers themselves or by subsequent scholarly commentators, to the social domain (i.e., to the social structures within which the environmental protection system is embedded and upon which it is dependent for its success and durability). Few have examined in detail the social foundations (values, traditions and practices) of the extensive legal and institutional framework for environmental protection that was put in place during the 1970s and 1980s, or the values and capabilities of the leaders and staff of regulatory agencies (the majority of whom were hired during the communist era), or the degree to which Poland's social structure may have been transformed under the communist rule in ways inimical to environmental protection. Some early analyses of the lingering societal effects of the Soviet era, for example, predicted a collective disregard for the law, which would clearly undermine efforts to improve environmental performance (Sztompka 1992; Wedel 1992; Tarkowska and Tarkowski 1989; Kolarska-Bobińska 1994).

Efforts to reform the environmental regulatory system have benefited from the overall success of Poland's economy in the 1990s and from the initial vitality and competency of the reform movement. New growth and investment serve to offset the economic effects associated with intensified enforcement of environmental regulations, while the post-revolutionary spirit readily accommodates new ideas and policy experimentation. There remains concern, however, as to whether the emergent environmental protection system is sufficiently sensitive to the economic fragility of many factories. The legislative reforms provided little statutory flexibility for licensing agencies or inspectors, even if factories were making good-faith efforts to improve environmental performance. Yet the structure of the institutional system in Poland put local and regional environmental inspectors in a good position to assess the scope of improvements that a factory might practically achieve and to identify any circumstances that present a real and immediate environmental or health threat. The 1999 decentralization of environmental protection in Poland only magnifies these strengths and weaknesses: the capacity for flexible facility-specific implementation and inadequate statutory flexibility to do so. The ongoing maturation of the environmental regulatory system, and the concomitant resistance to change and reform, might make it more difficult to make the necessary changes in the future.

The considerable reforms since 1989 notwithstanding, Poland has not leapfrogged, as some leaders of the early reforms had hoped, to a national environmental strategy aimed at a sustainable future. The decade of efforts has focused primarily on reducing pollution intensity of economic activities, especially in the energy and large manufacturing sectors, and end-of-pipe pollution control. While evidence is accumulating that these efforts have paid off, progress in ameliorating the environmental effects of continuing economic growth and consumerism in Poland has been modest (Millard 1998). Generally, the environment plays a minor role in the national political life and economic planning. For example, Karaczuń (1995, 1997) notes that National Policies for Transportation and for Energy Supply, issued by the government during 1994–1996, do not mention the environment (see also Stodulski 1999, and Kamieniecki and Kuspak 1998, for the critical analysis of the environmental strategy during the first post-communist decade).

Looking forward, the ability to respond flexibly to the circumstances of particular firms, industries and regions will be critical to the long-term success of the pollution control system in Poland. In general terms, existing research suggests two broad approaches to this challenge, one based upon increased use of market-based instruments, the other focusing upon the development of an information rich regulatory environment (Tietenberg and Wheeler 1998; Toman 1993). We pick up these and other issues in Chapters 4 and 5, which present our case studies of five recently privatized firms.

Chapter 3

Occupational Health and Safety System in Poland

The collapse of state-centered communist control in Poland in 1989 created a fundamental challenge to the country's occupational health and safety system, as the principle on which this system was founded—that the state represents workers' interests and owns the means for their protection—vanished along with the discredited political regime. As a consequence, policy makers seeking to improve occupational health and safety (OH&S) protection in Poland simultaneously face two daunting challenges. In the first instance the emergent new social order requires that employers, labor unions, workers and regulators and all other participants in the occupational protection system reexamine their roles, responsibilities and mutual relationships. At the same time there is an urgent need to address the failures and inadequacies of past policies and industrial practices, including such problems as institutionalized incentives for poor enforcement, a generally weak safety culture and a multitude of often unimplementable exposure standards.

These challenges are to be addressed under the difficult circumstances of dramatic socioeconomic change and in the face of multiple potentially competing objectives, such as reducing unemployment while also improving worker health and safety. Thus regulators are under pressure to improve working conditions while simultaneously considering the socioeconomic impacts of regulation on firms and communities. Labor unions, which are no longer the agents of the state, must reinvent themselves as independent advocates for good working conditions while trying to preserve the socioeconomic safety net created during real socialism. A rapidly growing private industrial sector is creating a new class of entrepreneurs with widely ranging degrees of experience in, and commitment to, occupational health and safety. The financial fragility of many newly privatized firms creates tension between occupational health and other business objectives. Finally, Poland's efforts to join the European Union call for harmonization of its system of occupational health and safety with the European approach.

Potentially, of course, the two dimensions of the challenge contain within them the seeds of a solution. Fundamental rethinking of the purpose and organization of occupational protection creates a window of opportunity to address and overcome problems inherent in the existing system. Depending on the particular situation, this might involve a radical break from the past, or a more incremental pattern of reform, keeping the best of the past and progressively eliminating weaknesses. As outlined in Chapter 1, however, such a positive outcome depends, among other conditions, upon all participants having a shared stake in the success of the system as whole, upon a capacity to learn and profit from change and upon the availability of appropriate policy instruments. Not all of these conditions were fully met with respect to Poland's system of occupational protection. As a consequence, a pattern of incremental reform during the 1990s has left unresolved a series of critical problems, including the need for specific procedures for balancing the many demands facing firms and regulators in Poland today.

This chapter describes the institutions and policies for occupational protection that have emerged in Poland since 1989. The analysis emphasizes both the recent reforms and the gradual evolution of the system over the past several decades.

LEGISLATIVE AND INSTITUTIONAL HISTORY

The legal framework for workplace protection in Poland has its roots in the pre-war period. The 1928 executive presidential order regarding occupational safety and health continued to serve as the legal basis for regulating working conditions until 1965. Several sets of regulations were issued during that period, including the comprehensive 1946 and 1959 ministerial orders that set the foundation for the development of regulatory infrastructure. Also, in 1952 the right of employees to a safe and healthy workplace was written into the Polish Constitution.

In 1965 (Public Law 1965) comprehensive legislation was adopted setting a legal framework for defining the respective roles of employers and employees and for defining the norms for achieving safe working conditions. In 1974 the Labor Code was issued, which incorporated most provisions of the 1965 labor law. The Code systematized and specified in significant detail numerous aspects of working conditions, including Section X (10) which addresses the issues of safety and health. The 1974 Labor Code became the actual legal framework for regulating working conditions in Poland. It accorded significant powers to the Ministry of Labor, Compensation and Social Issues (currently the Ministry of Labor and Social Policy) to modify workplace regulations through executive orders (Public Law 1974). The Labor Code has been modified numerous times since 1974, most recently in June 1996.

Apart from these developments in the legal domain, the post-war period witnessed the emergence of institutional infrastructure, much of which continues to this day. The Central Institute for Labor Protection and the Institute of Occupational Medicine were created in the 1950s to serve as the key research centers on occupational health and safety; the system of social labor inspectors working within enterprises was created at that time to represent interests of workers (their functions are discussed later); the ministry's enforcement arm in the form of labor inspectors (which dated to the pre-communist era) was separated from the ministry and incorporated into the functions of labor unions. Also during that period, the Office of the State Sanitary Inspector was created, taking over from the labor inspectors the functions pertaining directly to prevention of occupational disease.

The policies and institutions for occupational protection that emerged in Poland in the post-war period were almost exclusively the result of legislative and administrative initiatives. While there is evidence of concern expressed over the years among some scholars about working conditions and occupational diseases (Academy of Social Sciences and Institute for Study of Working Class 1987; Frąckiewicz 1989; Silesian Scientific Institute 1989), these intellectuals did not participate in creating the policies and institutions for occupational protection. The labor unions had somewhat greater input, however, they were not truly separate stakeholders during the Soviet-dominated period because of their close ideological identification with the state. As a result, the state was in effect the only "owner" of the system, both in the intellectual and practical sense: advocacy for workers' health and safety, critical assessments, innovative ideas and initiatives for reforms all originated mostly with the state bureaucracy.

BUREAUCRATIC APPARATUS FOR OCCUPATIONAL PROTECTION

The institutional arrangements described in this section were in effect until December 31, 1998. As a result of the decentralization of the governance system in Poland the institutional structure of the occupational protection system changed on January 1, 1999. These changes are described later in this chapter. For two reasons we describe in this section the pre-1999 regulatory structure. First, the case studies discussed in Chapters 4 and 5 were conducted in 1997 and can be understood only

Figure 3–1.

Entities Legally Mandated to Implement the Labor Code in a Workplace

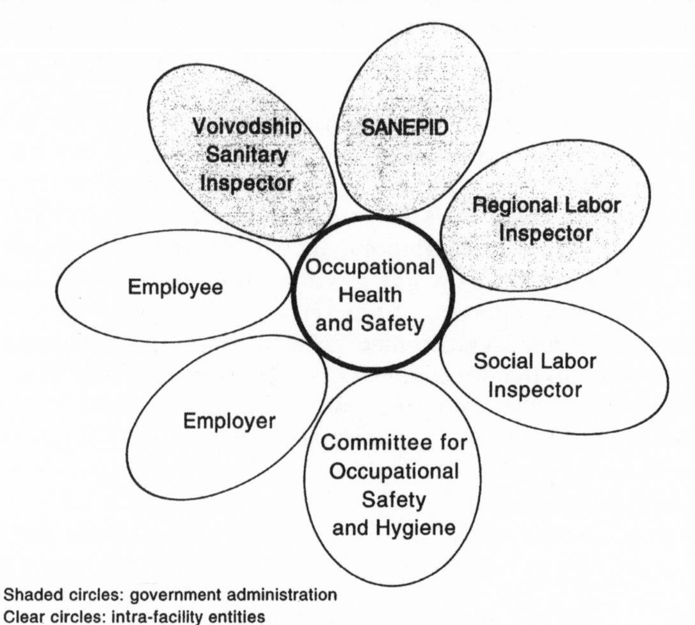

in the context of the institutional framework effective at that time. Second, the new arrangements have not significantly altered the types of decisions faced by regulators and firms alike.

The current occupational health and safety system in Poland gradually evolved over seven decades into an elaborate set of directives—the Labor Code—and a complex network of institutions that fall into two categories: formal governmental administration and an intra-facility system of various oversight units (Figure 3–1). Their functions and mutual relationships are described in the next two sections. As shown in Figure 3-2, the central governmental bureaucracy cuts across the legislative and executive branches of government and reaches out into regional and local organizations. The two branches of enforcement—the Labor Inspectorate and Sanitary Inspectorate—are collectively empowered with monitoring, inspection and enforcement functions (Public Law 1981 and Public Law 1985). The local organizations, Sanitary-Epidemiological Stations (SANEPIDs), are the state's furthest outposts into communities. With the exception of the Labor Protection Council, which is a new entity, all the institutions depicted in Figure 3–2 were created during the communist-dominated era in Poland.

The Ministry of Labor and Social Policy and the Ministry of Health and Welfare (thereafter referred to as Ministry of Labor and Ministry of

Figure 3–2.
Organization of Occupational Health and Safety in Poland (effective until December 31, 1998)

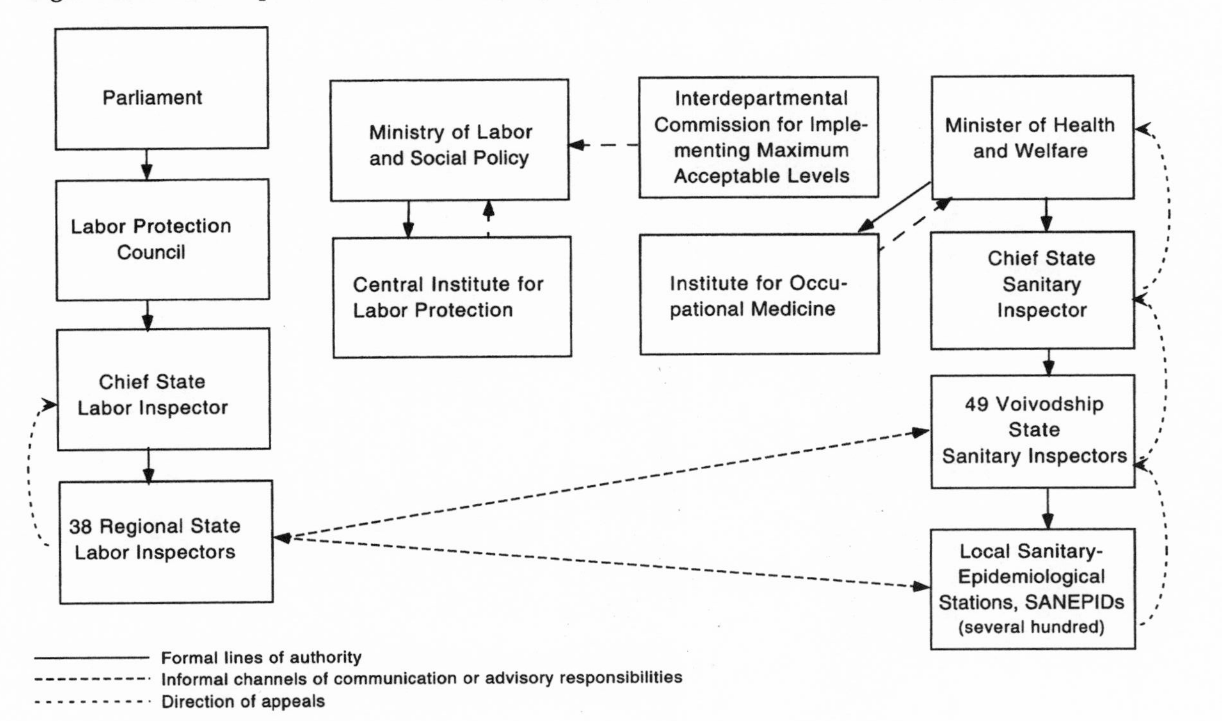

Health, respectively) are the primary policy-making organs. Notably, for each ministry occupational health and safety represents a relatively small part of their respective ranges of activities: the Ministry of Labor is primarily concerned with state employment policies and social benefits for workers, including job security and the compensation system, while the Ministry of Health is responsible for maintaining the national health care system.

In addition, the Ministry of Education plays a role in occupational health and safety through training and education of future engineers and other technical professionals. To preserve clarity, this and several other institutions with less central roles are omitted from Figure 3–2.

The Labor Inspectorate oversees a wide range of matters related to working conditions, including social benefits and general working conditions as well as safety and health. The power of the inspectorate has been significantly strengthened since 1989 through increased resources and administrative and fiscal independence from the policy-making ministry (since 1989 the chief labor inspector reports through the Labor Protection Council to the parliament). These functions are exercised through 38 regional labor inspectors who can impose fines, close work stations and production lines, suspend production and, in extreme cases, petition the chief labor inspector to close a factory. The fines for non-compliance have been increased in recent years by approximately an order of magnitude, to a level that is noticeable by medium-size enterprises (as of May 1996 the highest fine was approximately $2,000). Notably, the Labor Code specifically gives the regional labor inspectors a dual responsibility to enforce the law and to "give advice and technical assistance in the area of hazards to life of workers."

In contrast to the Ministry of Labor and the State Labor Inspectorate, the Ministry of Health focuses narrowly on the issue of prevention, recognition and treatment of occupational disease (not, for example, accidents or explosions). Its contribution to the Labor Code includes, among others, specifying the principles and frequency of monitoring indoor air at industrial enterprises, the medical examinations of workers, details on reporting of occupational disease and related health matters.

The Ministry of Health has an enforcement arm in the form of the State Sanitary Inspectorate (a marked organizational difference from the mutually independent Ministry of Labor and the State Labor Inspectorate). Like the State Labor Inspectorate, the State Sanitary Inspectorate implements its mandate through a network of regional offices. However, that network is far more extensive than that of the Labor Inspectorate, consisting of 49 voivodship sanitary inspectors and several hundred community-based SANEPIDs in individual *gminas*. In contrast to the regional labor inspectors and voivodship sanitary inspectors, local SANEPIDs have laboratory facilities for biological and chemical testing.

These laboratories are indispensable to the regional and voivodship labor and sanitary inspectors, respectively, for monitoring compliance with the Labor Code, especially with regard to compliance with occupational standards. This creates a permanent, if informal, line of communication on a regional and local level among these institutions (broken lines in Figure 3–2).

The history of the SANEPIDs is tightly linked to the history of public health, dating back to the nineteenth century and rooted in the prevention of outbreaks of infectious disease in communities. At that time these local health departments focused on clean water supply, sewage treatment, biological waste management, early detection and intervention in epidemics and other sanitary health matters. During the immediate postwar period, the SANEPIDs' major responsibilities within their communities were prevention, detection and response to outbreaks of infectious diseases and monitoring food safety. Occupational health was soon added to the SANEPIDs' range of responsibilities, partly for practical reasons (their testing laboratories allowed for monitoring levels of hazardous substances in workplaces) and partly for ideological reasons: the state's preoccupation with a worker as the foundation of the communist system. These functions have been retained after 1989.

The network of regional and local SANEPID stations, which are headed by medical and paramedical professionals, are closely rooted in community life. Employing largely women, these units often represent the first line of inquiry into the occupational health practices of a local manufacturer and the first line of problem identification and resolution.

The enforcement powers of the minister of health, the chief sanitary inspector and their regional and local representatives are much more limited than those of the chief labor inspector. They cannot close a factory for health or safety reasons and their fines for non-compliance are significantly lower than those of the State Labor Inspectorate. While they have the authority to suspend a production process if workers' health and lives are in jeopardy, they are not likely to exercise it in any but the most extreme circumstances. On the other hand, the Regional Labor Inspectors depend on local SANEPIDs for laboratory services and monitoring data and on voivodship sanitary inspectors for expertise and data on epidemiology of occupational disease. The sanitary and labor inspectors also jointly respond to the cases of non-compliance with occupational standards: while the labor inspector is most likely to issue fines, under the recent executive order the sanitary inspector determines the frequency of workplace monitoring, based on past performance record.

Despite their limited resources and enforcement powers, SANEPIDs have a unique role in affecting the working conditions within enterprises. Their frequent presence within a factory—to observe working conditions, monitor exposures to hazardous substances, examine health files and statistics, and respond to reported cases of occupational disease—is a powerful factor in modifying behaviors. Moreover, the SANEPIDs' uniquely narrow focus on workers' personal health allows them to give special emphasis to occupational health issues. SANEPIDs also have the flexibility to address situations that, while not strictly illegal, may increase risks of harm because their responsibilities are defined less formally than those of labor inspectors whose job is to strictly enforce the Labor Code. Finally, whenever necessary, SANEPIDs can readily mobilize regional labor inspectors and voivodship sanitary inspectors if the conditions in a plant are seriously compromised.

In short, the task of establishing and implementing policies for occupational protection in Poland is apportioned among three independent central institutions: two ministries and an independent enforcement agency. Their authority is exercised through a highly branched out system of regional and local administrators and inspectors. This administrative system is characterized by overlapping functions and responsibilities and mutual dependence among the local, regional and (to a lesser extent) central institutions and requires for its effective functioning a high level of cooperation and sharing of information among the participants. By all accounts, including our extensive interviews with the representatives of these institutions and case analyses of five firms, there indeed appears to be significant information exchange and cooperation among them, established over several decades of the system's existence.

THE LABOR CODE AND ORGANIZATION OF OCCUPATIONAL PROTECTION WITHIN A WORKPLACE

The intra-facility system of worker protection is rather elaborate. The Labor Code identifies several organizations, some in executive and others in the advisory roles, whose principal function is to assure safe and healthy working conditions in a workplace. These include: an occupational safety and health specialist employed by an enterprise (OSH specialist), an in-house testing laboratory, an occupational medicine physician (or a clinic), a social labor inspector, a Committee for Occupational Safety and Hygiene and a local labor organization. The regulations also provide detailed specifications for the state of machinery, technical devices and social and recreational areas and buildings (including such details as sizes of windows and showerheads) and requires periodic medical exams for employees. In addition, there are extensive requirements placed on employers for monitoring indoor air quality, human exposure to toxic agents, accident rates and occupational disease as well as for recordkeeping and reporting these statistics to the authorities.

Enterprises with over 100 employees are required to employ a full-time OSH specialist, while in smaller firms that individual may perform additional functions. Firms with fewer than 10 employees are exempt from this requirement.

A social labor inspector (Public Law 1983) is elected by employees to represent their interests before the employer. Since the elections are organized by facility-level labor organizations, the non-unionized enterprises have no social labor inspectors (although in such cases the employer is mandated to assure the presence of an elected representative of the workers). The regulations provide for certain powers and protections for social labor inspectors, such as protection from firing for two years after termination of the position and the mandate for an employer to implement inspectors' orders.

Each Committee for Occupational Safety and Hygiene is supposed to comprise the OSH specialist, occupational physician, social labor inspector and several elected worker representatives. Its primary function is to provide a forum for consultation between employees and employer, and to highlight areas needing improvements, although it also has the legal right (but not an obligation) to notify the state labor inspector of any violations of the Labor Code. The employer chairs and often controls the committee.

THE DECLINING ROLE OF LABOR UNIONS IN OCCUPATIONAL PROTECTION

During the communist period social labor inspectors were powerful advocates for workers' interests, backed by the authority of the state. It was not uncommon for social labor inspectors to confront plant managers and to close production lines if necessary. Over time, labor unions also assumed a prominent role in promoting OH&S protection. For example, regional Solidarity organizations provide training, develop guidelines for labor inspectors and hold the latter accountable for the activities of the Committees for Occupational Safety and Hygiene. Also, the facility-based unions organize the elections for the position of labor inspector.

The role of labor unions in the market economy has changed drastically since 1989. Once unambiguous proponents of strict occupational standards, in a free-market economy unions balance their advocacy for a healthy workplace with considerations of employment and compensation. It was plainly stated by one high level Solidarity official: "We have to worry about unemployment as much as about working conditions. In order to prevent layoffs, in many marginal enterprises we aim for gradual change and incremental improvements, such as rotation of workers assigned to particularly hazardous task, shortening of a workday, reliance on personal protective devices or extended time horizons for compliance." These sentiments were echoed by a top representative of the Federation of Labor Unions who told us that since the onset of the free-market economy, unions in factories have shown meager interest in working conditions, focusing instead on issues of employment and compensation.

The somewhat contradictory situation of Solidarity is illustrated by its position on the issue of compensation for harm. During the communist period the labor law provided for automatic monetary compensation to employees performing work in hazardous conditions exceeding occupational standards. The practice was so widely used, and considered so attractive by workers, that many deliberately sought out hazardous occupations and work assignments and opposed efforts to improve working conditions.

Since 1994 the legally guaranteed automatic compensation for working under harmful conditions has been eliminated, but not made illegal. Under the current law, employee benefits, including compensation for harm, are subject to negotiation between employer and employees and are included in collective labor agreements. The strongly entrenched culture of low regard for personal safety and the expectation of benefits often drive these negotiations in the direction of compensation and not elimination of hazardous conditions. Solidarity officially opposes any proposals to make illegal the system of extra pay for work in hazardous conditions. While the union prefers calling it "compensation for the burden of having to rely on personal protective devices," Solidarity views is as a well-justified tradeoff at the present time. Figure 3–3 shows that in 1996 approximately 34% of all workers in Poland collected some form of compensation for working under harmful conditions.

Along with the erosion of the unions' commitment to occupational health and safety came the erosion of the authority of the social labor inspector and the already weak Committee for Occupational Safety and Hygiene. As a result, the protection of an employee depends primarily on the attitude of an employer and vigilance of the state regulatory authorities.

OCCUPATIONAL EXPOSURE STANDARDS AND THEIR ENFORCEMENT

Occupational health standards are issued by the minister of labor upon a request by the "Interdepartmental Commission for Implementing Maximum Allowable Levels in the Workplace," which consists of members of parliament, technical experts and representatives of other ministries, employers and labor unions. The scientific basis of occupational standards are mostly developed by the Central Institute for Labor Protection, often with input from the Institute of Occupational Medicine.

Figure 3–3.
Proportion of Workers in Poland Who in 1996 Received Some Form of Compensation for Work in Hazardous Conditions (all economic sectors combined)

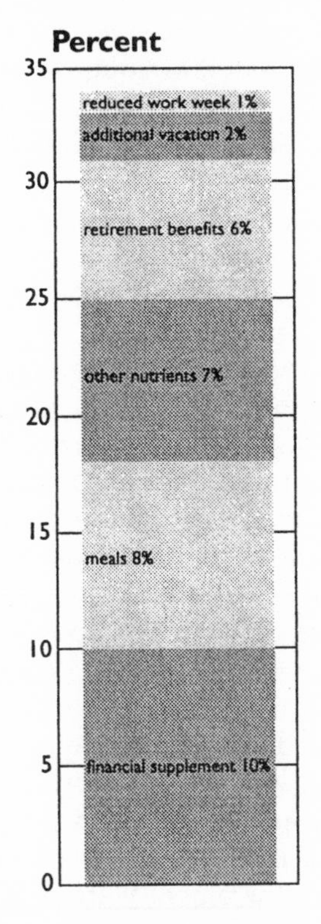

Source: Ministry of Labor 1996.

As of 1996 occupational standards have been issued for 303 chemical compounds and 13 classes of respirable dusts as well as for noise, vibrations and other hazardous agents and conditions (Benczek 1996), with many more in the planning stage. The Labor Code in Poland does not specify either the method for derivation of occupational standards or the factors that must be incorporated in considering the value of the standards (which is in stark contrast to the U.S. law). In fact, the Labor Code provides no guidance on how to balance health and safety goals with other social objectives in setting and implementing occupational standards.

Therefore, the parties involved in developing and issuing the standards—the Ministry of Labor, the Ministry of Health, the Interdepartmental Commission and the Central Institute for Labor Protection, which provides the scientific basis and justification for the standards (see Figure

3–2), have considerable discretion in determining the process and its underlying philosophy.

Occupational standards in Poland are based strictly on toxicological and health considerations, without allowance for technological feasibility, costs or compliance and enforcement issues. This approach, widely practiced in other Soviet satellite countries, was consistent with the state's official ideology of putting workers' interests above other interests and disallowing any explicit recognition of the need to balance health benefits with other considerations. Over the years, it produced a large number of strict occupational standards set at lower levels than those used in Western European countries and the United States. However, the implementation and enforcement of these ambitious standards lagged far behind their implicit promise.

Poor enforcement of the Labor Code and occupational standards was widespread during the communist era. For example, a controversial publications from the 1980s estimated that 7% of all employees in industry worked under conditions that violated existing occupational standards, and in some sectors of heavy industry this number reached 33% (Academy of Social Sciences 1987; Frąckiewicz 1989). Recent statistics suggest that the situation is not improving: 17% of workers in the manufacturing sectors were reported in 1997 to work under conditions harmful to health (State Labor Inspectorate 1998). Our own work on lung cancer epidemiology in Poland suggests that occupational exposures in the mining, metallurgy and energy generation sectors explained some of the observed elevation of lung cancer mortality in the most heavily industrialized regions of southwest Poland (Brown and Goble 1995; see also Indulski and Rolecki 1995).

The reasons for underenforcement were partly external to the system, affecting the occupational and environmental protection alike (Brown, Angel and Derr 1998). Thus the centrally planned economy included such non-negotiable realities as an emphasis on the development of energy- and materials-intensive industries; subordination of occupational protection to industrial production; full employment and economic growth; personal and economic disincentives for industrial managers to invest in occupational health and safety protection; and a general shortage of capital for occupational protection. In addition, the system for OH&S protection suffered from its own specific problems, such as unrealistically strict occupational standards and limited flexibility for inspectors to negotiate incremental compliance programs with employers; a low safety culture among employers and employees; and no tradition among workers of taking responsibility for their own health protection.

During the brief four decades after the war, Poland was transformed from a society in which 70% of population was rural into a society with 70% of population living in urban areas. A large proportion of industrial workers were first generation immigrants from rural areas: unskilled and unfamiliar with technology and with technological hazards. Occupational accidents and disregard for personal protection were high among that population.

The low safety culture was de facto encouraged by the state through the legally mandated system of automatic compensation and through salary increases, longer vacations, shortened workweek and other means for performing harmful work. Our case studies of five firms show how the bureaucratic apparatus responded to the reality of difficult-to-enforce standards and an unresponsive workforce: the labor inspectors practiced informal tradeoffs between occupational protection and other desirable objectives. These tradeoffs often took the form of extended compliance schedules, rotation of workers at hazardous job assignments, use of personal protective devices and others.

The principle of zero risk in derivation of occupational standards has been preserved in the post-1989 period of social reforms (Kowalski 1996). Today Polish occupational standards are more numerous and more stringent than their equivalents in Western Europe and the United States. A comparison of 1995 Polish standards with the U.S. ACGIH guidelines shows that 32% of Polish standards are lower by a factor of two or more than the corresponding ACGIH values and half of those differ by a factor of five or more. Only 10% of Polish standards are higher, usually within a factor of two. A comparison of the legally enforceable OSHA standards in the United States with the Polish standards shows that 64 % are higher than their corresponding counterparts in Poland by a factor of two or more, and half of those are greater by a factor of five or more. Several U.S. standards are two orders of magnitude greater than in Poland. Only one U.S. OSHA standard (for benzene) is lower that the Polish standard (1 and 10 mg/m^3, respectively, as an 8-hour average). The German and European standards resemble the U.S. ACGIH guidelines much more than the Polish standards (Benczek 1996).

The Central Institute for Labor Protection is a strong proponent of preserving the current practice. In an interview, the director of the institute, a three-decade veteran of the system, defined the issue in philosophical and pragmatic terms. Philosophically, stringent occupational standards are a declaration of values and declaration of long-term objectives. As a practical matter, faced with a weak personal safety culture in most industrial enterprises and a deeply entrenched practice of not complying with the standards, she views stringent occupational standards as a method of providing a margin of safety in workplaces that are not meeting them. "We know they will only partly comply with the standards so we might as well set them high to avoid health problems," she told us. The director also views the occupational standards as incentives for continuous efforts to improve the working conditions: a con-

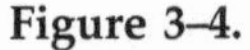

Figure 3–4.

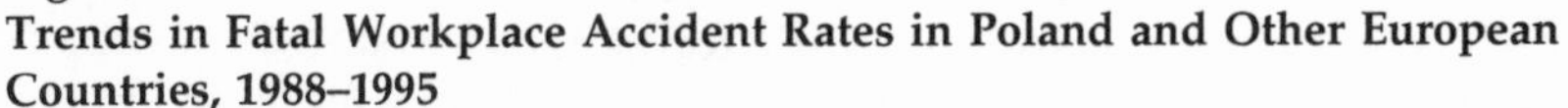

Trends in Fatal Workplace Accident Rates in Poland and Other European Countries, 1988–1995

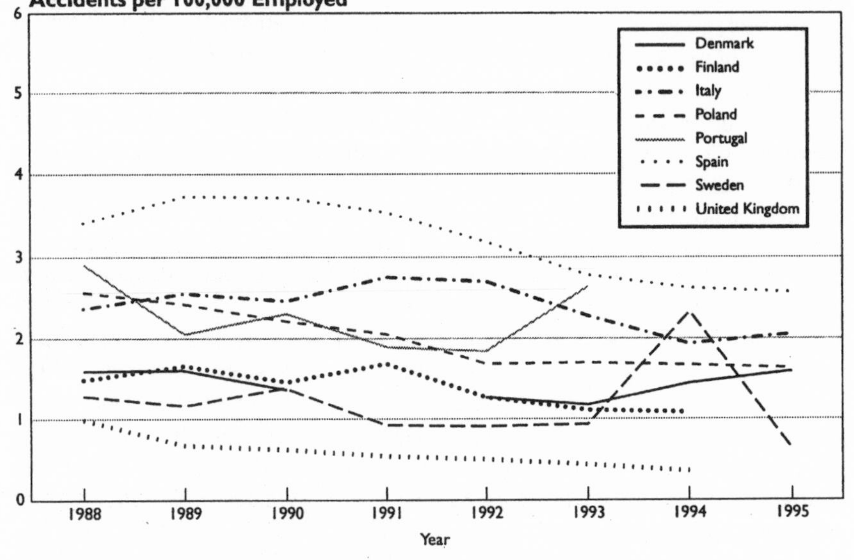

Source: World Health Organization 1996.

stant reminder to the employers, employees, labor inspectors, SANEPID inspectors, community labor inspectors and plant-level Committees for Occupational Safety and Hygiene that the goal has not been met and that further effort is needed. The Solidarity official we interviewed concurred with the views of the institute's director.

Despite the record of poor compliance, there is no evidence that the occupational protection system has experienced the type of massive performance failure that afflicted environmental protection during the Soviet era (Kabala 1985; Ember 1990; Institute of Environmental Protection 1990; World Resources Institute 1992). One of the voivodship chief sanitary inspectors put the issue in perspective during an interview by noting that the work-related accident rate dropped four-fold in Poland during the 1950–1990 period. Occupational statistics shown in Figure 3–4 show that Poland's record was comparable to that of other European countries at the end of the communist era, and that it has not deteriorated since. In fact, large portions of the fatalities are due to vehicular accidents and thus significantly elevate the mortality in the workplace (State Labor Inspectorate 1998, p. 153). In short, it appears that at least with regard to fatal accidents the key participants in occupational protection—employers, unions and government bureaucracy—share a common understanding that safety and health in the workplace cannot be allowed to deteriorate below some implicitly understood level of acceptability.

EXPERIMENTS WITH MARKET-BASED REGULATORY INSTRUMENTS

The use of economic tools in occupational protection has been the subject of many debates in Poland. One proposed plan, advanced by the Central Institute for Labor Protection (Central Institute for Labor Protection 1994; Rzepecki 1991), seeks to replace the current system of flat premium (45% of the payroll) paid by employers into the workers social insurance fund (which includes accident compensation) with a stratified system of premiums. The premiums would be stratified on two levels: First, the base for each industrial sector would be set according to the overall hazardousness of that industrial sector, based on the historical accident rate in that sector. Second, for individual firms the premiums would be adjusted upward or downward from the base, depending on their performance during the previous year relative to the average for that sector.

A formula proposed to calculate the magnitude of the performance-based adjustment considers rates of serious injury and occupational disease and the number of workers exposed to hazardous working conditions at that plant. Based on this formula, adjustments could range between 5% and 25% above or below the base rate. Notably, when asked for a rationale for setting a ceiling on the excess payment by poorly performing firms, the director of the Institute for Labor Protection noted that the proposed system is intended "to provide positive and negative incentives for behavior modification among employers and employees, rather than being a punitive instrument."

The appropriateness of this or other market-based incentive systems for the Polish context will be tested in the next several years: the system will become effective in year 2000.

FROM PATERNALISM TOWARDS SHARED RESPONSIBILITY

One of the significant changes introduced into the revised 1996 Labor Code is an attempt to reduce the paternalistic role of the state in regulating safety of the workplace by (1) increasing the discretionary powers and explicit obligations of employers and (2) increasing workers' participation in protecting their interests. The new regulations also eliminate any legal distinction between private employers and the state as an employer. Thus the code clearly places on the employer the ultimate obligation for protecting life and health of workers and for organizing the workplace in a way that would achieve these objectives. It also increases somewhat the powers of employees by giving them the right to refuse to work, without loss of pay, in any circumstances presenting imminent threat to health. Employers' obligation to establish the Committee for Occupational Safety and Hygiene in enterprises with more than 50 employees is also a new feature of the 1996 Labor Code.

Apportioning the responsibility for workplace health and safety among an employer, employees and the state bureaucracy would represent a major shift from the communist past when the state micromanaged every minute detail of life in a workplace, from specifying the shape and size of shower heads and water temperature to dictating the size and shape of windows. These reforms have been driven by several forces: the need to respond to the democratization of the society and to redefine the role of the state; attempts to contribute to building of a civil society where individuals and private institutions take more initiative and assume more responsibility for their actions; efforts to harmonize the Polish system with that of the European Union.

The explicit language of the Labor Code, which makes an employer ultimately responsible for the health and safety of employees, even in reference to business subsidiaries of larger firms and to joint ventures (where the legal accountability rests with the parent company) is only one manifestation of that effort. Another manifestation is the legally mandated requirement for workers' participation in the decisions related to occupational protection through the Committees for Occupational Safety and Hygiene and through the social labor inspectors. Introduction of market-based tools, as currently debated, is yet another manifestation of that trend.

These new developments are consistent with the general philosophy of occupational protection systems of the member countries of the European Union, but they urgently need further evolution before a system that is functionally equivalent to the European practice is established. For example, while the new Labor Code makes an employer responsible for training the workers and the OSH specialists, the philosophy and principles of that training are not defined. This is further exacerbated by the unraveling of the past organization of occupational training programs that were implemented by the Ministry of Education, often as part of professional training, at the institutions of higher education.

Also, the Polish system continues to rely heavily on detailed technical specifications for production technology, machinery, personal protection devices and organization of a workstation. The current reform movement within the government administration aims for replacing that practice with a combination of general principles and tools for risk assessment (to be performed by the employer), workers' training and informing workers about the risks.

The extent and timing of achieving these changes will clearly depend on the responsiveness of the corporate leaders, bureaucracies and workers to these new ways of conducting daily business of occupational protection administration. Our interviews and policy analysis suggest that this will be a slow process. First, the labor unions' disengagement from the occupational health issues is a serious barrier to change. Second, employees' interest in occupational health and safety lags far behind the matters of economic security. For example, while the number of inquiries by employees to the Labor Inspectorate is growing, less than 10% of the questions concern occupational health (the others concerning compensation, employment security or social services) (State Labor Inspectorate 1998). Additionally, policy leaders show little interest in invigorating the parties by, for example, engaging those who implement the Labor Code in the policy-making process.

CHANGES IN THE SYSTEM SINCE JANUARY 1999

As described in Chapter 2, on January 1, 1999, the governance system in Poland was decentralized. The number of voivodships is reduced from 49 to 16, and many local decisions are now delegated to the newly created elected county (*poviat*) governments led by *starostas*.

Figure 3–5 depicts the new institutional arrangements and lines of authority for occupational health and safety under the new administrative and political governance. Instead of 49, there are now 16 voivodship sanitary inspectorates offices and 33 regional sub-voivodship offices. In a major shift of the line of authority, the voivodship sanitary inspectors no longer report to the chief state sanitary inspector but are incorporated into the offices of *voivodas*. The heads of the sub-voivodship offices report to the voivodship sanitary inspectors. Another major change affects the local SANEPIDs. While some will probably report to voivodship sanitary inspectors (at the time of this writing, there was considerable uncertainty about this), most are now responsible to *starostas*, heads of elected country governments.

With *voivodas* and *starostas* controlling the regional and local branches of the Sanitary Inspectorate, the independence of this enforcement branch might potentially be in peril. However, several factors make this development unlikely. First, as shown in Figure 3–5, the appeals of the decisions of the regional and local authorities are heard by the higher instances within the enforcement branch, not by *voivodas* or *starostas*. Second, the reorganization does not affect the authority granted to regional and local sanitary inspectors by the Labor Code. Third, the Ministry of Health continues to be the source of budgetary allocations for SANEPIDs

Figure 3–5.

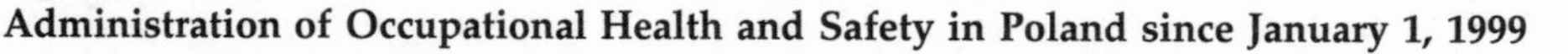

Administration of Occupational Health and Safety in Poland since January 1, 1999

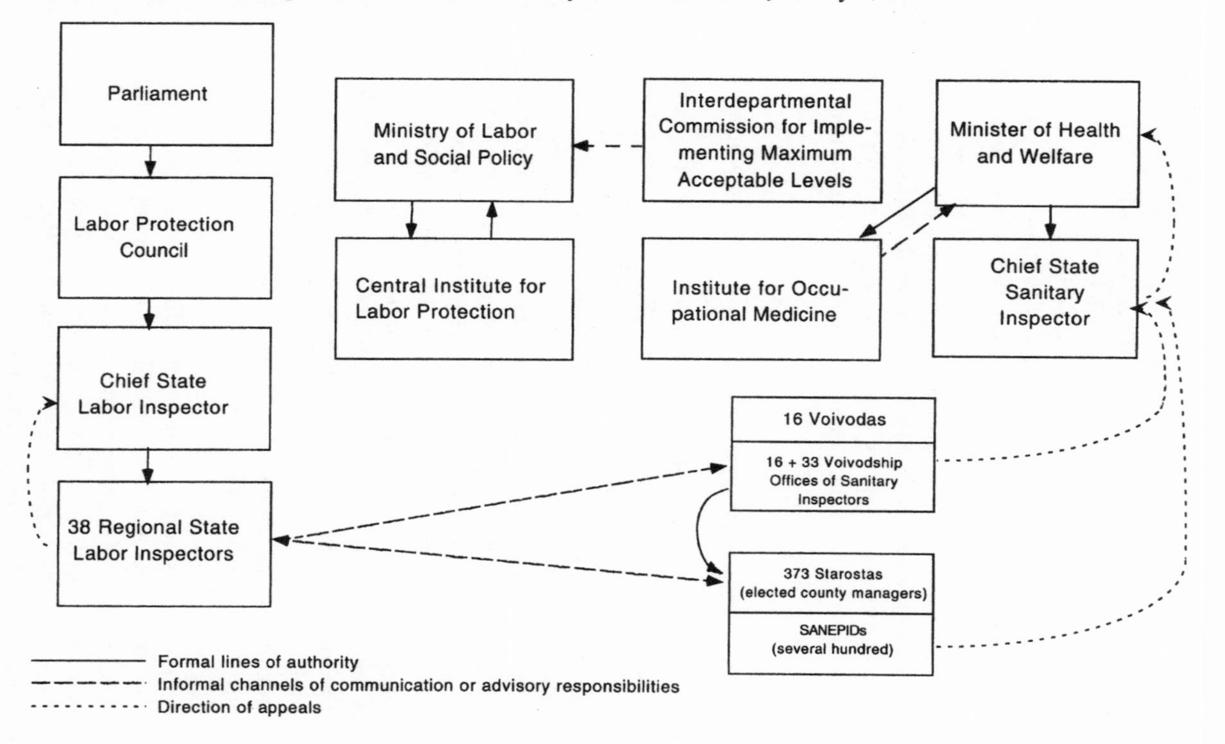

and regional sanitary inspectorates, which protects their independence. In addition, the new institutional arrangements are consistent with the established practices, such as the strong local presence of the community sanitary inspectorates, the key role of centrally funded SANEPIDs laboratories for both enforcement branches and the close cooperation between the labor inspectors and sanitary inspectors.

STRENGTHS OF THE SYSTEM AND ITS PRESSURE POINTS

The occupational health and safety system in Poland shares many structural characteristics with other OECD countries (see, for example, the comparative analysis of Belgium, Denmark, Germany, Japan, the Netherlands, Sweden, the United Kingdom and the United States by Prins et al. 1997). Like other OECD countries, Poland relies on government policies as the mainstay of occupational safety practices and enforcement. The right of employees to participate in development of employers' occupational safety and health policies is a shared feature of the Polish and other OECD systems (the United States is the exception). Poland also resembles Sweden and the Netherlands in its reliance on highly detailed regulations and norms, while its legal requirement that employers provide professional occupational health services to workers places Poland among the leaders in that area, along with Belgium, Japan and the Netherlands. Even the notable absence of economic incentives in Poland in the form of linking the accident insurance premiums to performance record is not an anomaly (such an incentive system is used by only four of the eight countries surveyed by Prins et al. [1997]).

Looking beyond these institutional and policy characteristics, we find that the prevailing occupational health and safety culture sets Poland apart from OECD economies. Here we refer to the weak safety culture among workers, who are not accustomed to taking a responsibility for their own safety, policy makers who have low expectations for safety performance in workplaces, labor unions that place safety and health in competition with other social objectives and an entrenched practice of non-compliance with occupational standards.

These attitudes are largely the legacy of the approach taken towards occupational protection during the communist period, including a highly paternalistic role of the state, a legalized system of working under hazardous and unhealthy conditions, very strict but difficult to enforce occupational standards and a state-sanctioned philosophy of putting industrial output and employment above all other national priorities. The widely shared post-1989 concerns about employment security, wages and the effects of maintaining high safety and health standards

Figure 3–6.
Activities of State Labor Inspectorate

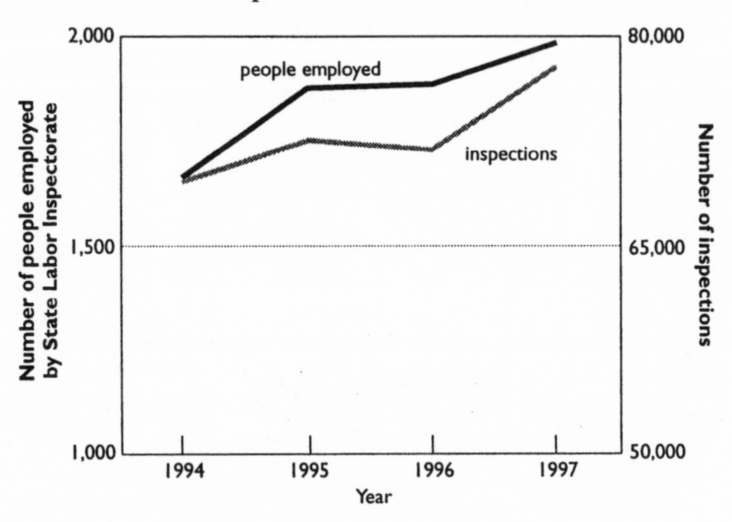

Sources: State Labor Inspectorate 1995, 1996, 1997, 1998.

in many financially fragile enterprises only further reinforce these attitudes.

The post-1989 reforms of the regulatory system for occupational health and safety in Poland have been decidedly incremental in nature. The system continues to comprise a complex set of institutions with overlapping functions, whose missions and mutual relationships are mostly the legacy of the communist past. The state continues to play a pivotal role in policy making and implementation.

Strengthening the enforcement branch is one of the most significant changes within the bureaucratic apparatus. On the facility side, most notable are recent efforts to increase the accountability of employers and to shift the responsibility for working conditions from the state to a shared arrangement involving the state, private employers and workers. The recently established Committees for Safety and Hygiene also hold promise. Studies in the United States have shown, for example, that workplace-based safety and health committees can be highly effective in lowering the rate of accidents and occupational disease (Wokutch 1990).

On the administrative side, elevating the stature and resources of the Labor Inspectorate (not affected by the January 1999 reorganization) and increasing the magnitude of non-compliance fines have been crucial. Figures 3–6 and 3–7 show that these efforts have paid off. The number of people employed by the Labor Inspectorate has been steadily growing (over 90% of the employment is in the regions where enforcement actions

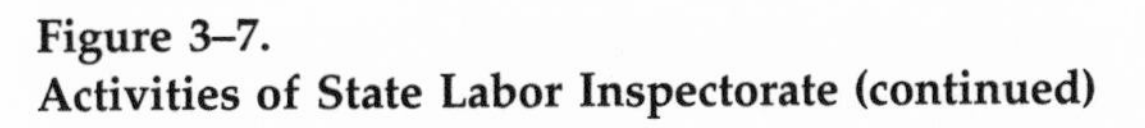

Figure 3–7.
Activities of State Labor Inspectorate (continued)

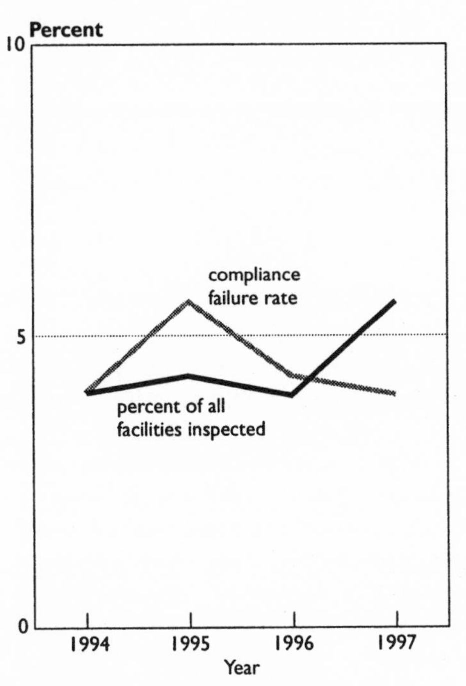

Sources: State Labor Inspectorate 1995, 1996, 1997, 1998.

take place) and so has the number of inspections. At less than 6%, the rate of non-compliance with the inspectorate's enforcement decisions is very low. The average record of 1.4 inspections per inspected facility per year (State Labor Inspectorate 1995–1998) suggests that the inspectorate focuses on facilities with particularly poor track records and that there is a follow up. In another illustration of the vitality of the enforcement branch, in 1998 the Sanitary Inspectorate for the Warsaw voivodship made 3,200 enforcement actions and in 20 instances closed production lines for health reasons (personal interview with director). The relatively low reported fraction of firms that are inspected annually (between 4% and 5%, according to Figure 3–7) applies to all firms and businesses registered with the Labor Inspectorate, including self-employed individuals and very small enterprises (less than 20 employees), and including all economic sectors. It, therefore, does not represent the situation within manufacturing facilities of our types. The statistics reported by the State Labor Inspectorate for manufacturing firms employing more than 20 persons give a very different picture. Thus, in 1997 the proportions of annual small, medium and large firms (21–100, 101–250, and over 250 employees) that were inspected were 36.5%, 52% and 60.1%, respectively. For all sizes combined the proportion was 40.3%.

For comparison, according to Garity and Shapiro (1993), in the United States only one in 25 workplaces have ever been visited by an OSHA inspector. Considering that in addition to the Labor Inspectorate the Sanitary Inspectorate also regularly visits the Polish firms, the picture of the oversight by the enforcement branch in Poland is very encouraging.

Notably, the post-1989 initiatives and ideas for reforms have come, as in the past, largely from within the legislature and state administration, primarily the Ministry of Labor and the Central Institute for Labor Protection. There is little evidence of a significant participation by independent intellectual elites within the academic community or non-governmental organizations. The labor unions, Solidarity and non-Solidarity, have shown limited interest in these matters, concentrating instead on such issues as employment security, compensations and social benefits, and in some cases have even opposed elimination of disincentives to improved safety culture in the workplace. Similarly, Western analysts, who since 1989 have engaged in a lively debate over the environmental policy reforms and performance in Poland and other former Soviet bloc countries, paid little attention to the occupational protection matters in that region.

This relative neglect of occupational health and safety can be partially explained by the lack of evidence that the occupational health protection failed during the communist period the way the environmental protection clearly has. For instance, Figure 3–4 shows that at the end of the communist era the fatal workplace accident rates in Poland were similar to those in other OECD economies. In addition, since protecting workers' health and safety was ideologically inseparable from other functions of the discredited political regime, the issue was never taken up during the Soviet era either by intellectual elites or the political opposition. Even at the height of its political prominence the Solidarity Union adopted the environment, not occupational protection, as one of its key themes for political struggle, even though logically the issues of worker health and safety were closer to its traditional domain. Finally, in the East and West alike the level of interest in occupational health has been for some time generally lower than in environmental protection issues. In short, it appears the government bureaucracy continues, as in the past, to be the main supporter and advocate of the occupational protection policies in Poland.

The incremental reforms rather than radical overhaul that characterize the post-1989 period offer many advantages. Building on a long tradition dating to the 1920s, they may gradually de-emphasize the shortcomings of the system while they build on its strengths, of which there are many:

the strong advocacy by the central institutions on behalf of workers, codified in the Labor Code; a wide network of regional and local inspection and enforcement institutions with considerable familiarity with the firms and with shared responsibilities; a long tradition of work towards prevention of occupational disease by the Sanitary Inspectorate network of institutions; a tradition of cooperation among agencies and employers; and rich accumulated experience in balancing competing social objectives.

The overlap in responsibilities among the multiple actors and dual responsibilities of the labor inspectors as enforcers of formal regulations and sources of advice and technical assistance can also be a source of strength to the system. It can contribute to building a sense of community and shared values, facilitate assessment of a problem and help in identifying negotiated solutions to problems involving conflicting or competing interests. Indeed, the community-based negotiated approach to management of technological hazards, which may be viewed as a manifestation of a working civil society within a uniquely Polish context, is a goal espoused by many policy analysts in the United States.

However, the post-Soviet changes within the Polish economy have been abrupt and profound, not incremental. The emergence of a strong private entrepreneurial class, growing competition in a free-market economy, economic fragility of many privatized enterprises, changing societal positions of unions and the state, scarcity of investment capital and the emergence of civil society have collectively created a very different social context for occupational health and safety protection.

When viewed in this light, the incrementality of change and certain features of the emergent system may lead to failures in protecting workers' interests. First, the facility-based system for occupational protection is fragile without the support of labor unions, possibly giving employers excessive freedom and latitude in managing workplace conditions. This is particularly problematic at the time of emergence of a new entrepreneurial class, often with little experience and motivation to manage occupational health and safety, leaving the advocacy for the workers largely to the state. While the Sanitary Inspectorate and its strong network of regional and local institutions concerned specifically with occupational disease may go a long way in compensating for this emerging impediment, they may not suffice in preventing some erosion in safety of workplaces. Based on the U.S. experience, where we have observed a declining power of organized labor since the 1970s, the ability of unions in Poland to affect the national occupational safety and health policies may continue to be limited as the country's economy shifts from manufacture- and infrastructure-based to service-based (Kuhn and Wooding 1997).

Second, in the absence of explicit recognition in official policies of the necessity for balancing competing objectives, and lacking guidance on how to do so, the system must most likely rely on an informal system of balancing through implementation. This is a precarious approach under the best of circumstances. For one thing the regional and local inspectors may be unprepared for the intensity of competition among multiple objectives in the increasingly competitive market-driven society. Their dual role as policemen and advisors may also conflict with the heightened expectations of strict enforcement of the Labor Code. In addition, the private sector is unlikely to continue to be as cooperative as it has been so far, especially when faced with shortage of capital and market pressures. In a spring 1999 interview the president of the Polish Association of Employers described a growing trend among employers to employ specialized legal and technical services for appealing administrative decisions. Finally, any community-based informal decision-making model of balancing competing objectives has unavoidable inherent weaknesses: it can easily breed excessive familiarity among the parties and increase the risk of corruption; it can also lead to diffusion of authority and responsibility and create disincentives for development of individual and institutional initiatives.

The third concern arises from the accumulated effect of chronic disregard for official standards and safety rules. The underlying assumption of the current reform program has been that, faced with changes in policies and policy instruments, and no longer pressed by systemic disincentives to compliance, the key actors would in fact change their behaviors, including respect for the official norms. This untested assumption is open to question because of the slow pace of any cultural change. This issue was discussed early on by several Polish scholars concerned with the overall attitude towards the rule of law among the Poles after several decades of cynical disregard (Sztompka 1992; Tarkowska and Tarkowski 1989; Obłoj and Kostera 1993). Institutional and policy analysis neither proves nor falsifies this hypothesis.

Finally, the narrow base of support for the occupational health and safety system among different societal actors is of concern. The gap left by the ongoing efforts to diminish the paternalistic role of the state makes the system particularly vulnerable to an organized challenge by the private sector, which is beginning to recognize its economic power. We have seen already extensive organizing within the energy, chemical an other key economic sectors in Poland around environmental regulations.

The private sector may, for example, choose to pressure the Ministry of Labor and Ministry of Health to relax occupational standards and other requirements on the employers or, as has been the case with most OSHA rulings in the United States, challenge the policies in court

(McGarity and Shapiro 1993). Should that happen, it is questionable whether the system is prepared to resist such pressure. For one thing, neither the currently practiced method for standards development nor its conceptual justification are codified in the law. In addition, the wide range of functions of both ministries, of which occupational protection is a relatively small element, will limit the degree of attention afforded to this matter. Even more importantly, the functions of the Ministry of Labor, which include matters of employment and social benefits to workers, create a powerful set of objectives directly competing with occupational protection objectives. The ministry would thus be a natural and susceptible target for any attempts to soften occupational protection policies in favor of other priorities.

These vulnerabilities will persist until other parties develop a sense of "ownership" in the system. One way of accomplishing this would be for the authorities to enlarge participation in policy making. This has not happened to far. During our interviews with employers and their organizations and with the enforcement personnel, we repeatedly heard reports of the absence of opportunities to participate in discussions of new policy initiatives or of unreasonable requests from the central authorities to comment on such initiatives within a few days.

Despite these concerns, we believe that there is basis for cautious optimism about the future performance of the occupational safety and health system in Poland. The system has numerous structural strengths, such as the dense network of institutions, its reach into communities, the intensive monitoring of firms' performance, the large number of occupational exposure standards and a vigorous and independent enforcement branch. The recently established Committees for Safety and Hygiene also hold promise. Studies in the United States spanning over two decades have shown that workplace-based safety and health committees can be highly effective in lowering the rate of accident and occupational disease (Wokutch 1990).

Another cause for cautious optimism derives from the very context in which the occupational safety and health in Poland must evolve. In the United States OSHA emerged in 1970 as a weak, underfunded agency with a vague vision and an ambiguous mandate, only to face a strong, well-organized and an hostile industrial sector. During the next three decades the agency was unable to alter that balance of power. Many of its subsequent failures can be attributed to these unfortunate beginnings and the subsequent history of reactive rather than proactive management (McGarity and Shapiro 1993; Noble 1997). In contrast, the Polish private sector is evolving in a world where the regulatory sector has already set the agenda and the rules and established its social status. Poland's close ties with its European neighbors and its desire to join the European Union also exerts a powerful influence.

In summary, analysis of the institutions and policies for occupational safety and health in Poland produces a mixed picture. The system features numerous strengths and, judging by the fatal accident statistics, has not suffered setbacks since 1989. The enforcement branch is robust and growing. On the other hand, there are justifiable concerns about effective policy implementation and the degree of support for the system among its diverse groups of participants.

Has the system been performing as suggested by the limited statistical data? What are the key variables explaining its performance during the recent years? How accurate are our observations about the strengths and weaknesses of the occupational safety and health system in Poland and about the implementation process? For answers to these and other questions we turn to the case studies.

Chapter 4

Case Studies of Five Firms

This chapter presents, in narrative form, the data collected in our five studies. It is a detailed description of the evolution of EH&S protection activities at five firms, as reflected in corporate, regulatory, legal and technological events and choices. So far as possible, this narrative focuses on the individual enterprises; interpretive and analytic remarks concerning the firms as a group are presented in Chapter 5.

These narratives are based on three principal types of data. First, hundreds of hours of structured interviews and informal conversations with key corporate officers, EH&S personnel, regulatory and enforcement officials, trade union representatives, national policy makers and other key actors. Second, an extensive paper trail, including corporate records and documents (many confidential), environmental impact assessment documents (*Operats*), permit applications, regulatory and monitoring records, orders for noncompliance fines, legal and administrative appeal documents, court decisions, national EH&S reports from governmental and other sources, memoranda from consultants and additional materials from municipal and other sources. Third, direct observations made by the research team (including senior CIOP personnel) during inspections of each of the five facilities.

The vital statistics of the three facilities studied are summarized in Table 4–1. These are medium-size enterprises, employing between 200 and 900 workers, that were privatized in the course of two years: 1994 and 1995. They are located in four different voivodships, but all are within 100 miles from Warsaw. The product lines vary widely among the firms, but they all handle materials with potential occupational and environmental hazards. At the same time these are not potentially heav-

Table 4–1.
Vital Statistics of Five Firms

	Drumet	Radom Leather Tannery	Fama	Raffil	Majewski Pencil Factory
City	Wloclawek	Radom	Wyszkow	Radom	Pruszkow
Voivodship	Wloclawek	Radom	Ostroleka	Radom	Warsaw
Year of Establishment	1895	1972	1963	1917	1894
Principal Products	Steel cables and wires for industry	Leather	Office and home furniture	Paints and finishes for industrial use	Pencils for office and consumer use
Principal Markets	Domestic and international	Domestic manufacturers of leather products	Domestic; former Soviet republics	Domestic	Domestic
Number of Employees in 1996	900	400	630	260	300
Year of Privatization	1994	1995	1994	1995	1995
Ownership	80% Private investment 20% Employees	85% National Investment Fund 15% Employees	Three Partners	100% Employees	65% Majewski Family 35% Employees
Labor Union?	Yes	Yes (Two)	No	Yes	Yes (Two)
Social Labor Inspector?	Yes	Yes	No	Yes	Yes
Committee for Occupational Health and Safety?	Yes	Yes	?	Yes	Yes

ily polluters such as would be paper mills, foundries, steel mills or manufacturers of synthetic chemicals.

Each facility was visited at least twice. Except in the case of the Radom Leather Tannery, we were able in each case to interview, inter alia, the firm's president (or CEO), the firm's technical or financial manager, the firm's occupational and environmental manager(s) and additional EH&S staff. At the Radom tannery (which was in the midst of a labor strike during our second visit) we were able to interview only the chief production manager. In every case we interviewed local and regional authorities responsible for environmental and occupational matters, including voivodship WWOS and PIOS officials, regional Sanitary Inspectorate personnel, and voivodship Labor Inspectorate personnel. Other actors were interviewed when warranted: in case three (Fama), local SANEPID personnel; in case two (the tannery), the local mayor and many of his administrative staff.

Some interviews with government officials were conducted in group settings. An unexpected benefit of this activity was that it brought together environmental and occupational officials who ordinarily have little (if any) contact; the resulting exchanges of ideas—especially regarding the relative merits of cooperative and confrontational approaches to regulatory enforcement—were extremely informative. The group interviews also gave us a close look at the working relationships between agencies which are encouraged, but not mandated, to cooperate: WWOS licensing officials, PIOS enforcement officials and SANEPID public health personnel.

Despite these many independent lines of investigation, we were not always able to secure sufficient information to reconstruct all the main EH&S themes with confidence. In cases where the information was too fragmentary to support any kind of cross-confirmation, we have preferred silence to speculation. Thus, in the case of the Radom Leather Tannery, we discuss liquid and solid wastes (including the "toxic lagoon"), but not air emissions; in the case of the Radom paint factory, we discuss air emissions but not liquid or solid waste.

CASE 1: STEEL CABLES MANUFACTURER DRUMET

Overview

Drumet is located in Włocławek, a voivodship city of 120,000, 100 miles northwest of Warsaw. Drumet is one of four major employers in Włocławek, the other three being a fertilizer manufacturer (over 4,000 employees), a paint manufacturer (approximately 2,000 employees) and a paper mill (1,000 employees). The enterprise was established in 1895 as a producer of wire fencing and soon expanded its production to steel cables for the mining and construction industries. Steel wires and cables are now Drumet's major product. Before 1939 the company had two different owners. After a temporary takeover by German authorities during the war, the enterprise was nationalized. By 1948 Drumet was again fully operational, employing 320 workers. In the 1950s the management began a program of technological modernization which culminated, in 1973, in relocating the factory from its original premises to a new thoroughly modern facility in a different neighborhood of Włocławek. By all accounts Drumet flourished during that period: much of its equipment was imported from Western Europe; it had substantial export markets in both the East and the West; and by 1975 its workforce exceeded 2,000 employees.

When the socialist regime fell in 1989, Drumet was a financially healthy enterprise with a well-defined product line and market. It was unencumbered by debt and well poised to compete in the emerging market economy. During that year the top management was replaced by a new team who have continued to the present and who led the company through privatization in 1994. Eighty percent of Drumet's stock is owned by a Polish investment firm in Poznán and 20% by the employees. Privatization was followed by a significant reduction in the size of the workforce, achieved, according to the president, through attrition and retirements. The current workforce consists of close to 900.

The Drumet facility abuts a densely populated residential area of 25,000 people. The proximity of factory to employees' neighborhoods is a legacy of communist industrial and urban policies: the major manufacturer provided affordable neighborhood housing for its employees, schools, energy, health centers, recreational centers and other foci of community life. While schools, medical care and other social amenities have since been decoupled from private business activities in Poland, Drumet has retained some elements of that culture. The company is presently refurbishing a vacation center for its employees and company literature emphasizes Drumet's stability, loyalty to its workers, and fidelity to the values of its original founders. Despite such traditional elements, Drumet is a thoroughly modern enterprise with a complex management structure, specialized departmental functions (including units for environmental and occupational health) and a world-class approach to marketing and corporate image building.

Drumet's chief products are steel cables for the fishing, petroleum, mining, construction, automotive and other industries, as well as wires for furniture manufacturing and other uses. Between 1992 and 1995 total production increased by 41% (from 19,900 to 28,103 tons per year). The proportion of exports also increased, from 5.8% in 1993 to 18.9% during the first eight months of 1995. The company's foreign markets include nine countries in Western Europe, the United States, Egypt and South

Africa. In spite of increasing competition from Asian manufacturers of steel lines, Drumet has maintained and expanded its foreign markets.

The company takes particular pride in the high quality of its products, which carry formal certification from several institutes of standards in Poland and in Western Europe. The company's publications also highlight its links with research institutions in Poland. Among its numerous awards and achievements are a gold medal from the 1992 International Fair for Innovation in Brussels, a Gold Statue awarded to leaders of Polish business by the Business Center Club and a First Prize in a national competition for the safest employer, sponsored by the National Labor Inspectorate. Clearly, technological modernization, product quality and successful competition on international markets were central Drumet strategies long before privatization.

The occupational protection unit in Drumet dates back several decades and currently consists of two employees. The director is a 25-year veteran in the company. He reports to the president, along with managers of several departments related to human resources, exports and imports, chief counsel and others. During the 1980s, the occupational hygiene department was expanded to include environmental pollution control specialists. The environmental protection group soon outnumbered the occupational protection group, and in 1991 the two groups were separated. Currently, the Department of Environmental Protection has 20 employees, mostly engineers and other technical specialists. The director, a 15-year veteran at Drumet, reports to one of two vice presidents, along with the directors of several technical and production departments. The company also has a social labor inspector and a Committee for Occupational Safety and Hygiene.

Management of Environmental Hazards

The manufacturing processes responsible for most air emissions, water discharges, and waste generation are: high temperature treatment of steel (consisting of treatment with hydrochloric acid and molten zinc); painting operations; and mechanical processing of steel wires and cables. For decades hydrogen chloride, lead oxide, volatile organic compounds and aluminum oxide were the primary air pollutants. The principal wastewater pollutants included lead oxides, zinc, other metals, cooling agents and various hydrocarbons. In 1994 lead was eliminated from the manufacturing process and thus from air and water discharges.

The company is required to obtain permits from WWOS for air emissions, solid waste disposal and chemical waste disposal. Wastewater disposal is governed by a contract with the municipal sewage treatment works. The history of these permits, particularly the air permits, illustrates the nature of interaction between the facility and the environmental regulatory authorities.

Solid and Liquid Waste Disposal

Solid waste generated at Drumet consists mainly of fabrics and wood-shavings (some contaminated with oil) and old equipment and construction materials (some contaminated with glues and adhesives). It is disposed of at a sanitary landfill in another city. Chemical waste consists of sludge from the on-site water treatment facility, metal shavings and sediments from metal treatment baths. The sludge has a high metal content. The wastes are managed in a variety of ways, including on-site treatment, recovery and reuse, resale as raw material to other industries and shipment off-site to a municipal hazardous waste disposal facility. In 1994 the top layer of soil throughout much of the premises was removed and shipped off as part of a general environmental upgrading at the enterprise. As shown in Table 4–2, during the 1991–1995 period, no fines were imposed on Drumet for waste handling.

The firm's records indicate that the permit for solid waste disposal at a sanitary landfill was issued by WWOS only 10 days after the submission of the application in November of 1994. The same speed characterizes the response by the voivodship environmental protection authorities to Drumet's request for permission to send oil-contaminated solid wastes for incineration at the town power plant: the permit follows two weeks after the submission of the May 1995 application. According to Drumet management, metal shavings, sediments from metal treatment baths and sludge from wastewater treatment are shipped off-site by a licensed hauler. We found no paper trail to indicate how the enterprise and the environmental authorities arrived at this solution. After pretreatment on site wastewater from the facility is drained into the municipal sewage system. The 1992 and 1995 contracts between Drumet and the municipal sewage works are highly detailed: allowable levels of over two dozens organic and inorganic water contaminants (including several heavy metals, ether extractable organics and volatile substances) are precisely specified. The allowable concentrations of most contaminants covered by the 1992 contract are taken from the guidelines provided by the Ministry of Environmental protection. Drumet, however, negotiated with the sewage works a zinc standard that is more than an order of magnitude less stringent than these guidelines. This was possible, the environmental director explained, because Drumet is the only significant zinc source loading the Włocławek sewer system, and its total zinc discharges do not threaten the municipality's compliance with national discharge standards for surface waters. (Drumet's ability to negotiate a relaxed zinc standard in the absence of other zinc sources was balanced by the au-

Table 4–2.
Environmental Fees and Fines Paid by Drumet between 1991 and 1995 (in thousands of złoty)

	Fees (x 10^5)			Fines			
Year	Water Use and Wastewater	Industrial and Solid Waste	Air Emissions	Wastewater	Air Emissions	Industrial and Solid Waste	Total fees and fines relative to total sales
1991	731	19	14	92	2	0	2.4%
1992	501	43	18	18	0	0	1.5%
1993	403	42	33	0	0	0	0.9%
1994	363	131	35	0	0	0	0.7%
1995	583	20	44	0	0	0	0.7%

thorities' ability to impose extraordinarily strict hydrochloride standard in the presence of other significant hydrochloride sources: see below.) Several other water quality standards were relaxed in the 1992 contract, relative to its predecessor. The changes benefited both parties: because it was compliant with the new standards, Drumet replaced fines (which are paid from net profits) with fees (which are built into operating costs), while the municipality obtained increased income from Drumet.

In the early 1980s the company launched a long-term project to dramatically upgrade its on-site wastewater treatment facility. The modernization was completed, in stages, by the mid 1990s. Drumet now operates a completely automated facility that can be operated by one person and, according to the environmental director, produces technologic grade water. As shown in Table 4–2, since 1992 Drumet eliminated all fines for wastewater discharges.

Air Emissions

Since the 1970s airborne emissions of hydrochloride gas have been the greatest environmental hazard at Drumet. The manufacturing process uses approximately 80,000 pounds of hydrochloride per week (2,000 tons per year), and until recently much of this was lost to the atmosphere from point sources or fugitive emissions. Fugitive emissions, although not subject to standard air permit regulations, can have a significant effect on local air quality. Drumet's hydrochloride emissions had been a source of tension between the enterprise and the environmental authorities for years. According to the voivodship environmental authorities we interviewed the short-term ambient air standard for hydrochloride has been frequently exceeded over the years in the area around Drumet, and they attributed this to Drumet. Figure 4–1, based on monitoring conducted by PIOS, shows that the 24-hour standard was exceeded slightly in 1991 in one location in the housing development adjacent to the facility. (We have no data to assess the degree of compliance with the ambient standard prior to 1991.)

We reviewed the hydrochloride emissions paper trail between Drumet and WWOS back to 1986. Hydrochloride emissions were clearly contention between the enterprise and the authorities for at least a decade. The air permits issued by WWOS in 1986 and 1987 set the total emission rate for three point sources of hydrochloride gas at 0.4604 kg/hour (0.372 + 0.062 + 0.0267).

But within a year WWOS revised its decision. In a new permit issued in 1988, it required Drumet to reduce total hydrochloride emissions from the three sources by a factor of 7.6 (for the three individual point sources, by factors of 11, 6 and 2, respectively), to 0.0606 kg/hour. According to WWOS's *technological feasibility* analysis, measurements taken at the fac-

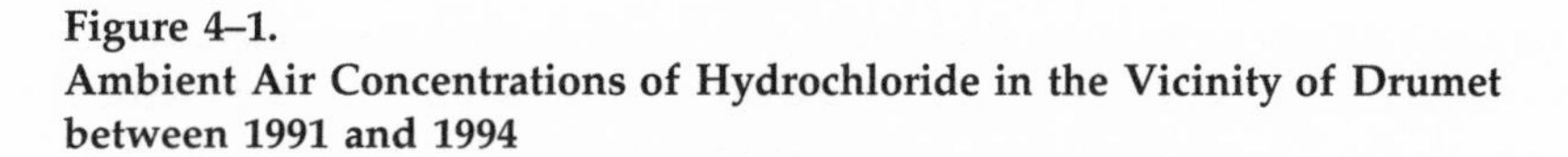

Figure 4–1.
Ambient Air Concentrations of Hydrochloride in the Vicinity of Drumet between 1991 and 1994

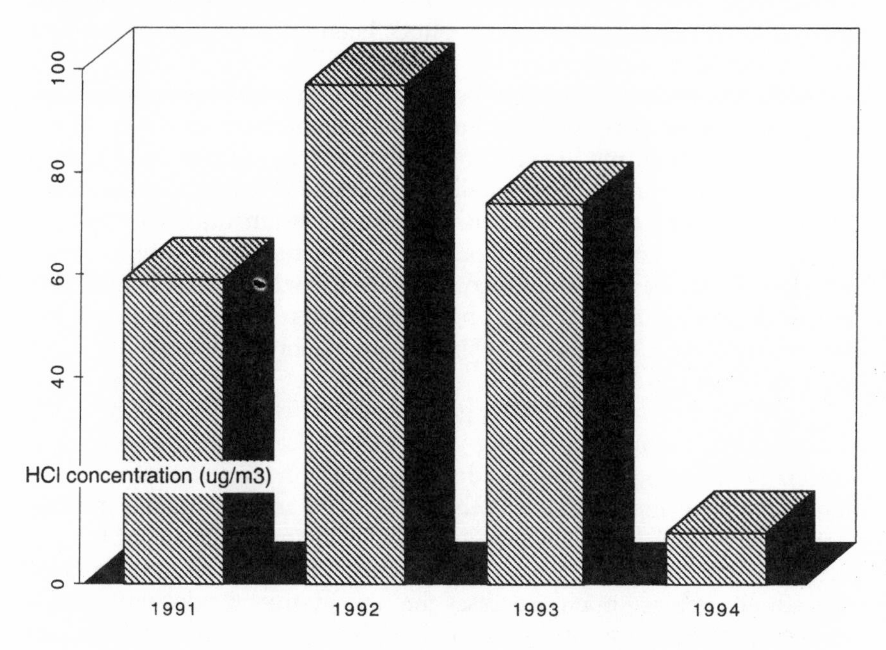

tory had shown that available technology could achieve significantly lower emission levels than those specified in the 1987 permit; hence, the new emission rates were well within the capacity of the technology available at Drumet at the time.

Within days of its receipt Drumet appealed WWOS's 1988 air permit decision to the Ministry of Environment and Natural Resources. Drumet presented two arguments. Using a *technological feasibility* argument, management pointed out that the low emissions achieved in the past already represented the best attainable results and that under average operating conditions the equipment could not possibly achieve the hourly emissions rates required by WWOS. Using a *public health* argument, management pointed out that the Drumet facility was already fully in compliance with the ambient air standard for hydrochloric gas. Pointing to its good will and earnest past efforts to protect the environment, the company argued that "it took a great deal of effort on our part to achieve average annual ambient concentrations far below the legal standard . . . and therefore we ask for positive response to the request."

The ministry's response, issued only a few days after the appeal was received, denied Drumet's request to roll back the new requirements. The ministry, which clearly used input from the voivodship office, reiterated WWOS's judgment that the new emission rates were technologically feasible. Most notably, the ministry also pointed to the issue which, as not subject to formal permitting, had not been formally addressed by either party: the adverse effects on public health of the extensive fugitive hydrochloride emissions from the factory.

Another round of exchanges between Drumet and the ministry followed in quick succession in August 1989, without gains for the enterprise. Drumet attempted to respond to the issue of fugitive emissions by appealing to the ambient air quality standard, which, according to the enterprise, would be met even if the total emission rates were almost doubled. It also complained about the authorities' insensitivity to the enterprise's extensive investments in improved environmental performance. The ministry rejected both arguments, and a final appeal by Drumet to WWOS, in June of 1990, was predictably rejected by the voivodship authorities.

Drumet's ability to predict the government's behavior was now made clear. Even as it was conducting its formal appeals, the enterprise was simultaneously taking steps to adopt a new technology that would permanently solve the emissions problem. Corporate documents indicate that by early May 1987—as though anticipating the intentions of WWOS to tighten the emission standard—the enterprise had begun a program to gradually replace open processes that use hydrochlorides with closed processes. The program was initiated almost immediately, and by May 1996 the most-used seven of 13 production lines had been replaced, at a cost of $400,000 each. The current plan calls for the replacement of all 13 lines by 2003. The new, closed-process system produces no air emissions and includes technology that purifies, recycles and reuses all captured hydrochloric acid. In fact, Drumet now sells its hydrochloride-purification services to other enterprises.

Given the earlier loss of appeals by Drumet, the last as recently as 1990, it might seem surprising that the 1993 air permit issued by WWOS allows for total point-source hydrochloride emission rates that are almost 75% higher than those specified in the hotly contested 1988 permit. In part this may be due to the new system of required Air Quality Assessment Documents that, since 1990, must accompany all applications for air permits. This system forces both sides to justify their proposals and decisions on technical grounds and, while making enterprises more accountable and open about their environmental releases, also puts limits on discretionary behavior by environmental authorities.

The 1993 permit was issued only days after Drumet submitted to WWOS the required Air Quality Assessment Document and air permit application. Both had been prepared on behalf of the company by a ministry-certified expert. The permit used the report's recommendations without any changes. But if WWOS seemed to relent on point-source rates, it could afford to do so: the recommendations in the assessment document were based on the newly installed closed-process and recycling technologies—and these had significantly reduced not only the point emissions but also the pernicious fugitive emissions which were so irritating to the authorities.

The 1993 Air Quality Assessment Document also addresses emissions of nitrogen dioxide, carbon monoxide gas and aluminum oxide particles. Lead, which was to be phased out the following year, is excluded from the analysis. The analysis uses the standard methodology for calculating air emission rates: the allowable contribution of Drumet to the air concentrations of the above pollutants is a residual calculated by subtracting the background concentrations in the area from the ambient air standard. Since the background concentrations—determined by the voivodship's sanitary inspector in Włocławek—are quite high in that area (ranging from 15% of the maximum allowable concentrations for hydrochloride to 75% for particulate matter), relatively small increments are left to Drumet. Based on the thus-calculated incremental ambient concentrations, the document recommends emission rates for each of the individual regulated pollutants.

The three year air permit issued by WWOS only two weeks later again used—without change—the emission rates recommended in the Air Quality Assessment Document. The company did not challenge the regulatory decision.

Shortly after WWOS issued the 1993 air permit, the Voivodship Environmental Protection Inspectorate (PIOS), following its own annual schedule for 1994, conducted extensive air monitoring for hydrochloride in the neighborhood surrounding the Drumet factory. Although it focused on three locations with the highest expected impacts, all ambient concentrations were far below the 1990 ambient air standards of 100 ug/cu.m. hourly and 20 ug/cu.m. annually: measured levels were between 2% and 20% of the 24-hour and between 1% and 2% of the annual standard, respectively. Furthermore, the ambient concentrations were far lower than the corresponding concentrations recorded by PIOS in 1991, 1992 and 1993.

The results of the 1994 monitoring are surprising. The 1993 WWOS air permit assumed that background hydrochloride concentrations equaled 15% of the ambient standard. Even assuming a *zero* contribution from Drumet, measured concentrations should have been at least that high. But with one exception, they were not. It would appear that the emission rates proposed in the Air Quality Assessment Document (a document prepared by a consultant hired by Drumet) allowed a significant margin of safety by overestimating both existing background hydrochloride levels and the potential impact of Drumet's hydrochloride emissions. WWOS awareness of these overestimations may help to explain its adoption, without modification, of the levels recommended in the document. But our interviews suggest that two other factors also influenced WWOS: First, an aggressive and well-informed concern about technological feasibility, not just meeting the legally enforceable ambient air standard; and second, a more accommodating attitude towards the enterprise because of its progress in solving the problem of fugitive emissions.

As shown in Table 4–2, 1991 was the last year Drumet faced non-compliance fines for hydrochloride emissions. This is consistent with the PIOS's air monitoring results shown in Figure 4–1.

The 1994 Air Quality Assessment Document also addressed emissions of volatile organic compounds (VOCs) from Drumet's painting operation. This situation was particularly difficult for Drumet because background levels due to other large manufacturers in Włocławek were already reaching allowable ambient concentrations. WWOS again adopted the Air Quality Assessment Document's recommendations without change and issued the air permit on the same day the document was received. Drumet was required to increase its stack height and to strictly control VOC emission rates in order not to exceed its allowable contribution to local air concentrations. Viewing stack extensions and other changes as expensive short-term solutions at best the environmental director chose to switch to water-soluble paints. Since January 1995, Drumet has used water-based paints exclusively.

In summary, continuous improvements in environmental performance at Drumet have been tightly coupled with technological modernization and with long-range financial planning based on a keen sense of future regulatory decisions and a desire to avoid investments in merely temporary solutions. Regulatory pressure, in the form of air emission permits, wastewater standards and non-compliance fines, provided important incentives for these improvements. WWOS's demands regarding air emissions were largely determined by technological feasibility, but sought both to comply with air standards and to protect public health. The successive positions taken by WWOS suggest that it will push very hard for technological improvements when it judges that the enterprise is financially able to implement them.

From the company's perspective, the technological modernization imposed by WWOS and the ministry, though initially resented and challenged, was quickly integrated into the company's plan to become a leader among Polish private industries. A forward-looking and dedicated environmental manager, supported by a president who viewed strong environmental performance as a powerful business asset, enabled the enterprise to achieve these forward-looking changes. Indeed, Drumet's strong environmental performance increased the market value of the enterprise and was an important bargaining chip during privatization negotiations.

The legal formalities of privatization in 1995 were hardly noticeable from the environmental protection perspective. Most of the changes aimed at pollution reduction had been initiated in the early 1990s, and all were on the way to completion. Continuity of professional personnel was of far greater significance. Most of the environmental protection personnel, including the environmental director, were retained through privatization. This group of individuals had molded the environment- and modernization-friendly culture of the enterprise over two decades. That future is now supported and reinforced by top management.

The environmental director's personal philosophy, his ability to make his case to corporate management and his clear knowledge of both the regulatory environment and its key regional officials have all been factors in Drumet's success. His personal philosophy includes a belief in long-range planning (evident in his consistent rejection of stop-gap solutions), a strong sense of community (he is active in several local volunteer programs), a strong feeling of moral responsibility for the well-being of the 25,000 people living around the factory ("I have 25,000 people out there to answer to" was his comment to us) and a very deliberate strategy of engaging regulatory authorities as peers, if not as partners.

Corporate management uses the firm's reputation for environmental leadership as a marketing tool. In 1994 the enterprise signed a Declaration of Cleaner Production with the Ministry of Environmental and the Ministry of Industry. And a preliminary audit for ISO 9000 certification was conducted in June 1996.

Although, as shown in Table 4–2, Drumet had eliminated all environmental non-compliance fines by 1993, the enterprise continues to pay substantial fees for air and water discharges, waste generation and water use. In 1995, these fees represented 0.7% of its annual sales.

Management of Occupational Hazards

Occupational safety issues have a long and troubled history at Drumet. And unlike environmental management, where improvements began in the early 1980s, occupational safety and health progress did not receive serious attention until approximately 1990.

Drumet's main occupational hazards include noise, airborne dust (including metal dust) and mechanical injuries. Until 1994 lead exposure was a serious concern at some workstations. And as discussed earlier, hydrochloride fumes were a long-standing and serious hazard for some workers.

While lead was still used in the manufacturing process, the local SANEPID visited Drumet frequently, and the company had a history of non-compliance with the occupational lead standard. As required by law, the enterprise conducted regular surveillance of workers' blood and urine lead levels but, according to the voivodship labor inspector, "the records released by the enterprise were open to questions." Nonetheless, there is no record of a major confrontation between the authorities and the management over the lead problem.

Meeting the occupational standard for noise has always been problematic. Recent capital investments in new closed equipment have improved the situation, but the quarterly monitoring reports submitted by the enterprise to the voivodship sanitary inspector continue to show occasional cases of non-compliance at some workstations.

The company has developed an elaborate training system aimed at the prevention of accidents and occupational disease. As described in a voluminous manual (the most recent version issued in 1986), the system calls for training courses, periodic examinations of workers for knowledge of safety rules and financial incentives for job safety performance. Nonetheless, in practice, the system was clearly ineffective in preventing accidents and serious injuries. During the 1970s and 1980s, the enterprise reported up to 150 serious accidental injuries per year for a workforce with fewer than 2,000 production employees, and one Labor Inspectorate inspection cited over 270 violations of the Labor Code.

Throughout the 1970s and 1980s the Labor Inspectorate kept a watchful eye on Drumet, making frequent unannounced inspections. Nevertheless, we found no record of significant fines or of any threat to close production lines. According to the occupational safety manager, the management of the facility began to pay attention to safety issues during the 1980s, and the injury rate then began to decline. As recently as 1990, however, accidents at Drumet still accounted for 50% of all eye injury admissions to the local hospital, leading some local citizens to call Drumet "the generator of blind people."

According to the representatives of voivodship Labor Inspectorate and SANEPID, major changes in Drumet's attitude towards occupational health and safety took place when new management took over in 1989. The accident statistics, shown in Figure 4–2, confirm this observation. The 1995 total accident rate (for accidents associated with a loss of one or more workdays) was 21 per 1,000—only a third of earlier levels, but still an order of magnitude higher than the national average for the metal industry (approximately 2 per 1,000). During our inspection of the factory we saw several production workers not wearing safety glasses. Others, in high noise areas, were carrying, but not wearing, ear protectors. Attitudes towards personal safety still have room for improvement.

Occupational health has improved more rapidly than occupational safety, in part, perhaps, because of its synergistic relationship to environmental protection. The new management made fundamental changes in the occupational health area. First, it addressed the lead issue by upgrading the on-site testing laboratory and performing an extensive anal-

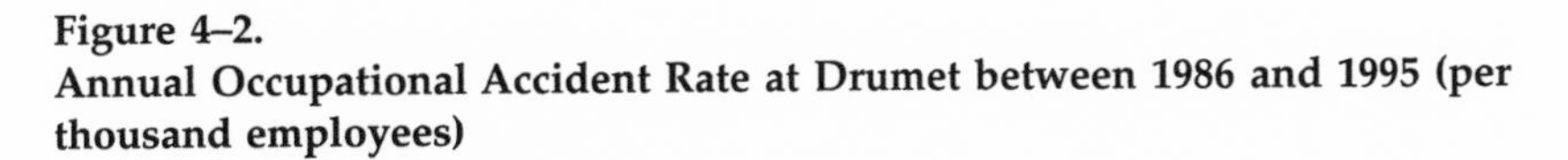

Figure 4–2.
Annual Occupational Accident Rate at Drumet between 1986 and 1995 (per thousand employees)

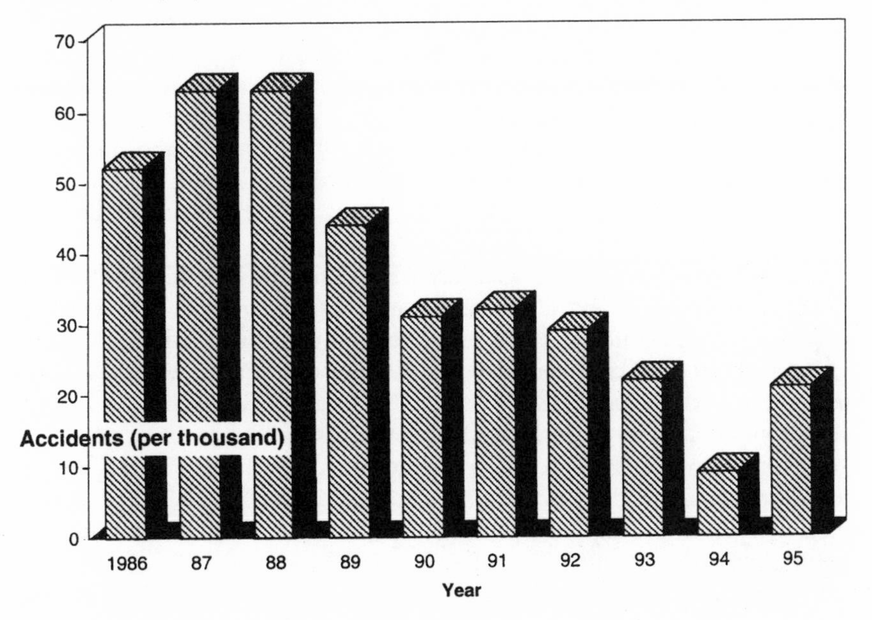

ysis of workers' lead exposure; finally it eliminated lead from the manufacturing process altogether. Second, it drastically reduced occupational hydrochloride exposure by modernizing equipment, as described earlier. Third, the shift to water-soluble paints also reduced exposures to volatile organic solvents.

In 1992 Drumet contracted with the highly regarded Institute of Occupational Medicine in Łodź to perform a comprehensive analysis of indoor air concentrations of lead and the body burden of lead among 104 workers whose jobs brought them in contact with lead. The study found that only one employee was exposed to airborne lead levels exceeding the occupational lead standard and that 87 of 104 workers had non-detectable blood lead levels. The remaining 27 workers had levels between 10 and 32 ug/dl, all well below the then acceptable blood lead level of 60 ug/dl. The institute found Drumet fully in compliance with the occupational lead standard. The 1993 inspection report by the voivodship Labor Inspectorate confirms the Institute of Occupational Medicine findings and also shows compliance with occupational zinc and hydrochloride standards between 1991 and 1993. According to the report, all but one workstation were in compliance with both standards throughout that period. The voivodship sanitary inspector also visited

Drumet in 1995 to monitor compliance with occupational standards for toxic substances and found no violations.

Following the Institute of Occupational Medicine's favorable report on lead exposure, Drumet management eliminated "hazards compensation" for workers coming in contact with lead. This compensation included shortened workdays (from 8 to 6 hours), higher salaries and additional vacation days. Workers, supported by their trade unions, vigorously resisted the loss of benefits and tried to enlist the regional labor inspector for their cause. The representative of the Labor Inspectorate told us, "For two years we continued receiving complaints from the workers, and the union even contemplated legal action to restore the benefits!"

Neither the unions, the social labor inspector, nor the Committee for Occupational Safety and Hygiene played a significant role in restructuring the safety and health culture at Drumet. Every individual we interviewed about this case remarked that the local union had little or no interest in occupational health issues. For example, according to the occupational health manager, the social labor inspector has taken no action at Drumet since 1990. And the position taken by the unions regarding the elimination of the lead hazard compensation system suggests that their interest in preserving benefits greatly exceeds their interest in eliminating hazards.

It appears that regulatory agencies played only a supportive role in the turnaround at Drumet as well. But because the timing of the change in the management and the reform of the enforcement bureaucracy coincide, it is impossible to know what role the authorities *might* have played at Drumet, had the new management continued to be lax on occupational issues after 1989. The voivodship labor inspector commented that "because of the positive attitude of the new management, we were able to shift from our earlier confrontational attitude to our current cooperative relationship." This comment evidences the agency's preference for a cooperative, negotiated approach to problem solving—a preference echoed by other actors as well. But it also confirms our impression that management was the driving force for change.

Today, Drumet has a well-established reputation for strong occupational health performance and an improving record for occupational safety. In 1994 the company received a safety award from the Central Labor Inspectorate for meeting every occupational standard with an ample margin of safety. According to the representatives of SANEPID and Labor Inspectorate we interviewed, three factors were pivotal in effecting this dramatic change: the attitude of the top management, the company's financial and professional resources and the nature of the Drumet workforce. The president sent an unambiguous message that occupational safety and health mattered, and backed it up with a performance-based incentive system and a strongly centralized (some used the term "authoritarian") management system. Our interviews revealed that disregard for internal safety policies now carries a swift penalty at Drumet. Drumet's financial health and EH&S expertise were also crucial: together, they made it possible to design and implement corporate policies requiring major capital investments. Finally, the workforce at Drumet, largely continuing from the pre-1989 period, is disciplined, highly skilled and well educated—good material for the introduction of modern management techniques.

In sum, the dramatic changes in the occupational performance at Drumet have been mostly the result of management efforts. It is impossible to say how deeply rooted the new safety culture is at Drumet. Clearly, the goal of ISO 14000 certification is a powerful motivator for top management to improve the company's performance. The stability of the enterprise will help strengthen its corporate culture.

Summary

Drumet is a success story. In less than a decade the company achieved major improvements in environmental and occupational performance, established itself as a leader in EH&S and maintained both market share and product quality. In the environmental arena, government authorities were involved in forcing technological changes that brought environmental improvements. But there is abundant evidence that the company's technological initiative and shrewd assessment of the future market value of being "green" were also key factors.

In the occupational protection arena the evidence points more directly to the leading role of the company's management in achieving improvements. Legally mandated workers' organizations showed little interest over the years in reducing the accident rate or eliminating chronic health hazards. Occupational protection authorities kept a close watch on the facility but tolerated the unacceptable rate of occupational accidents until 1990. While they clearly welcomed the post-1989 improvements, there is little evidence of their contribution to the changes. Key changes in environmental and occupational performance, including accident rates and investments in cleaner technologies, can all be traced to the early 1990s, when new management took over the facility. Privatization, in 1994, appears to have played no significant part in these developments.

Our study of this case revealed a complex, multidimensional relationship between Drumet and the environmental authorities, a relationship consisting of a mix of command-and-control and cooperation. At first glance WWOS's approach to hydrochloride emissions suggests an arm's-length relationship with little room for negotiations. However, a deeper analysis of WWOS's decisions, based on both interviews and the documentary paper trail, reveals that the agency pressed the firm to precisely and only the level it *judged technologically and financially feasible*. Moreover, it is clear that WWOS had the information and expertise to gauge this feasibility quite accurately. In reality, throughout the years covered by our case analysis, the voivodship regulatory agency was skillfully balancing health and environmental gains against costs to the company and (indirectly) the community.

This balancing took on a new dimension in 1990 with the introduction of the requirement that enterprises applying for an air permit submit an Air Quality Assessment Document. The Air Quality Assessment Documents submitted to WWOS by Drumet were prepared by a consulting firm under the direction of a ministry-certified expert. It is clear that the analyst was thoroughly familiar with both the technical capabilities of Drumet and the objectives of the permitting agency. The fact that all permits were issued within days of application and that the permits use language and conditions identical to those in the applications suggests that the expert served as a trusted mediator between the two parties. Clearly, the content of the application and the permit were not a surprise to either party, suggesting a rich flow of information between the parties.

The relationship between Drumet and the municipal authorities (over disposal of wastewater) is also characterized by complex negotiations. The allowable concentrations of certain contaminants were raised in the 1992 sewage works contract, to the benefit of both parties and without (according to the representatives we interviewed) adverse environmental impact.

On the occupational protection side the relationship between the enterprise and the authorities was distinctly skewed in favor of the enterprise under the centralized pre-1989 system. State enterprises had the latitude to act according to their own priorities, and production goals clearly trumped health and safety. Because the onset of occupational health and safety improvements at Drumet coincided with the reform and strengthening of the Labor Inspectorate, we cannot assess how this newly empowered agency would have responded during the 1990s had no improvements taken place at Drumet.

CASE 2: RADOM LEATHER TANNERY

Overview

Radom Leather Tannery is situated in Radom, a voivodship city of 230,000 located approximately 70 miles south of Warsaw. Traditionally a center of light manufacturing and agriculture, the Radom voivodship and its capital city have declined economically and currently face a 20% unemployment rate. Leather tanning and shoe manufacturing have for a century been among the dominant local manufacturing industries. Despite some decline in that sector, the Radom voivodship still hosts over 150 tanneries (37 in the city) and several leather goods manufacturers. Some of the smaller tanneries are relatively new, and employ former workers from faltering larger enterprises.

Since the early 1990s the Radom city government has been very active in environmental affairs. In the 1994–1996 period it participated in an ambitious cooperative project with the U.S. EPA, the Institute for Sustainable Communities in Vermont, and the Polish Institute for Sustainable Development to assess environmental conditions in the city and the surrounding region and to develop a strategic plan for future environmental improvements. The effort identified groundwater depletion, groundwater contamination, and sewage treatment—each a major impact of the leather tanning process—as three of the area's top five environmental problems.

Radom Leather Tannery was established in 1972 as part of a large state-owned leather manufacturing conglomerate which produced shoes and other goods for both national and international markets. In 1990 the three tanning plants (the largest in Radom) were separated from the larger enterprise, probably to enhance the larger enterprises attractiveness to potential investors. The newly independent tanning entity had 1,100 employees, including 900 at the Radom plant. By 1996 employment at the Radom Tannery had declined to 430, largely due to the general decline in shoe manufacturing nationwide. Current management hoped to reduce the workforce by another 10–20% by 1997.

In March 1995 ownership of the Radom Tannery was assumed by and divided between the state-owned National Investment Fund (85%) and the employees (15%). The tannery is starved for capital and struggling to maintain markets for its products. It is unable to invest in improvements to its outmoded infrastructure, its manufacturing process or its environmental and occupational safety systems. Its future now seems to depend on finding a private investor willing to assume the majority ownership of the enterprise and provide a large capital infusion.

The current director of personnel, who was our main contact at the firm, it is a long-time employee of the original concern and, for several years, of the (now) Radom Tannery. After spending several years of directing the Environmental Department, her responsibilities have been expanded to include production management and social services. In the course of our lengthy interviews we discovered a person of considerable experience, managerial skill, strategic vision and political acumen. It was clear, both from her extensive comments and from her multiple responsibilities at the tannery, that all her decisions related to the environment, occupational health, personnel and production are very consciously interconnected. The Radom Tannery employs both an occupational health specialist and environmental management specialist, both of whom report to the director of personnel. Unfortunately, despite repeated requests from senior CIOP officials, the president of the enterprise was not willing to meet with us.

On visiting the site we found a series of large manufacturing halls, clearly underutilized, poorly lit and accessible only through long windowless corridors reminiscent of nineteenth century factories. The need for maintenance and modernization was evident at every step.

Radom Tannery presents a wide range of environmental and occupational hazards common to the tanning industry. The process consumes a very large amount of water and generates a large volume of chemically contaminated wastewater. Among the occupationally and environmentally hazardous chemicals are hydrogen sulfide gas, sulfuric acid, sulfates, sulfites, ammonia, trivalent chromium and organic solvents. Respirable dust, noise and mechanical hazards are significant occupational concerns. In addition, chromium-laden leather shavings are a formidable waste disposal problem.

Under current environmental regulations, the tannery must obtain multiple permits from the voivodship in order to operate. These include permits for the use of water, the discharge of industrial effluents into surface waters, air emissions, disposal of solid waste at the municipal landfill, storage of solid waste on site, and disposal of industrial sludge in a nearby surface lagoon. In addition, a permit from the municipal sewage treatment works is required for wastewater discharges into city sewers. All these permits carry fees based on the quantity and nature of the material, and fines for non-compliance. The annual fees and fines paid by the tannery are summarized in Table 4–3 and represent approximately 5% of all fixed operating costs.

Management of Environmental Hazards

The Radom Leather Tannery faces a formidable spectrum of environmental problems. These include: an outdated wastewater treatment plant that does not meet required performance standards; a dangerous surface lagoon containing a quarter century's chemical and biological wastes; and on-site storage of chromium-laden leather shavings. The facility is also the object of persistent neighborhood complaints about odors from hydrogen sulfide discharges into city sewers. The relationship between the facility and the authorities over these environmental problems reveals much about the environmental management system currently shaping up in Poland.

Wastewater

The tannery uses 2,000 cubic meters of water per day and discharges most of it into the municipal sewers as wastewater. A renewable contract

Table 4–3.
Environmental Fees and Fines Paid by Radom Leather Tannery between 1990 and 1995 (in thousands of złoty)

Year	Fees			Fines		
	Wastewater	Industrial and Solid Waste	Air Emissions	Wastewater	Air Emissions	Industrial and Solid Waste
1990	4	30.1	no data	0	0	0
1991	55.2	128.7	3.5	94.5	0	0
1992	147.1	121.5	0.3	0[2]	0	0
1993	312.9	no data	0.6	74.3	0	0
1994	353.7	0[1]	0.7	0[2]	2.3	0
1995	536.3	0[1]	1.8	153.1	1.0	0

[1]Decision pending.
[2]According to the director, the lack of fines was an oversight on the part of municipal sewage works.

with municipal sewage works specifies allowable levels of contaminants in the wastewater. The tannery's wastewater treatment facility, built in 1972, was grossly inadequate from its inception. During the past two years the company has installed a chromium recovery unit, which has significantly reduced chromium concentrations in the wastewater.

Our case analysis took a close look at contract negotiations between the Radom Tannery and the Radom Municipal Sewage Works for the 1990–1994 period. The contract was signed in 1991 after a year of negotiations over such issues as the number of water quality parameters that would be subject to monitoring, the monitoring frequency and who would bear the costs of analysis. In each round of negotiations the sewage works made its proposals and the tannery accepted or rejected various conditions. On its part the tannery skillfully (but selectively) invoked regulatory provisions and coolly accused the city, which had not completed construction of a new treatment plant, of "attempting to pass on the consequences [of this inaction] to the industrial enterprises in the city of Radom." This accusation came from an enterprise that for over two decades has discharged chromium-laden wastewater directly into the municipal sewers!

Eight of the 11 water quality standards in the 1991 contract reflect official values adopted by the Ministry of the Environment. Three, dealing with contaminants that would be a major concern to any tannery, do not: the chromium standard is an order of magnitude higher, the sulfates standard is three times higher and the total suspended solids standard is two times higher. Plainly, the final terms of the 1991 contract were crafted to accommodate the special needs of the Radom Tannery.

Negotiations over a new contract were initiated in 1994. The sewage works notified the tannery that it intended to raise the water quality standards for nine contaminants covered by the 1991 contract and to impose an additional standard for detergents. The changes were drastic, reducing allowable concentrations by factors between two and 16. The tannery immediately rejected the new rules as a violation of a provision in the *existing* contract that "all changes in the quantity and quality of wastewater are subject to agreement by both parties." The final contract, signed five months later, rolled all the water quality standards back to approximately the 1991 level. The additional parameter was eliminated as well.

Even with these more tolerant water quality standards, the tannery bears a considerable financial burden for disposal of its wastewater. As shown in Table 4–3, its municipal sewage fees have, since 1991, been increasing at a rate outpacing the inflation rate. Production during the same period was declining. Furthermore, in 1995 the tannery paid noncompliance fines (primarily for exceeding the chromium and suspended solids standards) equal to 29% of its fees.

Radom Tannery's wastewater is a major burden on the neighboring housing development of 20,000 people because of the presence of foul-smelling hydrogen sulfide. This volatile substance, produced through decomposition of biological wastes, is not covered by the wastewater contract and thus is not subject to punitive additions by the city. WWOS has no jurisdiction in this matter. The PIOS official we interviewed reported on her agency's extensive monitoring of indoor and outdoor air for hydrogen sulfide, but in the absence of air standard violations, her agency had no basis for action. Thus the neighborhood currently has little legal recourse against this daily nuisance.

Hazardous Waste Lagoon

The outdoor storage site for liquid chemical waste is the most pressing environmental hazard associated with the tannery. The waste originates from the wastewater treatment plant as 98% aqueous sludge. The installation of the chromium recovery unit in 1994 has not reduced the volume of post-treatment sludge, with all its other contaminants. The liquid does not meet the wastewater standards and disposal into the sewers is not an option.

Built in 1979 as a temporary measure, the lagoon is an unlined and uncovered 0.6 acre pit in the sandy ground, 1.5 to 2 meters deep and surrounded by a retaining wall of soil, 4–6 meters high and 3.5 meters wide. Approximately 140,000 tons of liquid waste have accumulated there since 1979, containing chromium, other metals, solvents and biological agents. The lagoon is practically full and in constant danger of overflowing. (It has overflowed several times in the past.) As far back as 1986, studies showed that leachate from the lagoon was contaminating the underground aquifer with inorganic ions and organic solvents. This is a particularly serious concern because the city obtains its water from deep wells. We found no evidence of discussions about remediating or restoring the site, either on the part of the municipal or the voivodship authorities.

In the 1980s the tannery proposed to develop a new liquid waste disposal site and obtained tentative approval from WWOS. However, because the proposal was inconsistent with the local land development plan, administrative courts struck it down. Since then the tannery has been repeatedly pressing the city council to alter the land development plan, but even with support from the local government the chances of a new facility are rapidly declining in the face of public opposition. As of 1997 the prospects for including a new disposal site in the local land development plan seem practically nil.

In the meantime the voivodship has been struggling with the difficult decision to close the existing lagoon, fully aware that, in the absence of viable disposal alternative, closing the lagoon is tantamount to closing the tannery. The correspondence between the tannery and WWOS illustrates the dilemma: in recognition of the economic difficulties at the tannery, and despite incoming technical reports of further deterioration of the groundwater quality, WWOS postponed the inevitable decisions by granting temporary extensions in 1990 and 1991. These, in turn, only generated more requests for extensions.

In early 1992 WWOS refused to grant another extension. Radom Tannery promptly appealed the decision to the Ministry of the Environment, astutely portraying itself as a victim of both the city's failure to approve construction of a new disposal site and the National Environmental Fund's refusal to provide funds for waste treatment technology. It highlighted its recent installation of the chromium removal unit as evidence of its good faith effort to solve the problem. The tannery also pointed a finger at WWOS for its failure to develop a long-term plan for hazardous waste management in the Radom voivodship, and finally, the tannery emphasized the unemployment that would be generated by closing the facility.

The appeal was successful, and the tannery received a permit to use the lagoon until the end of 1992. But in January 1993, when the tannery asked for another extension, the Radom *voivoda* issued a categorical denial. In justifying the decision the *voivoda* cited both the results of groundwater quality analysis showing continuing deterioration and an expert technical analysis (prepared a few months earlier) which concluded that the lagoon's retaining walls were at a constant risk of a breach.

Undeterred, the tannery again appealed to the ministry. Losing the appeal this time, it turned in 1995 to the highest administrative court. The tannery's appeal defined the issues very narrowly, focusing only on the safety of the lagoon's retaining walls, and skillfully exploiting the carelessness with which WWOS and the ministry had used technical arguments in their earlier decision. First, according to the tannery, the environmental authorities misinterpreted the technical analysis of the safety of the retaining walls, which was the basis of their decision. Second, despite a request by the tannery, they did not admit into consideration the opinion of another expert who had earlier concluded that the use of the lagoon through 1996 would present no threat to the environment if certain specific precautions were taken. Hence, according to the tannery, the authorities' decision was invalid on two counts: it was based on false technical premises and on flawed administrative procedures.

In a decision issued only a week after receiving the tannery's appeal, the court sided with the tannery. Responding to the specific argument presented, the court agreed that (1) neither technical assessment explicitly stated that further use of the lagoon beyond 1992 would create a risk of breaching the retaining wall, provided that certain technical improvements are made, and (2) the authorities violated procedural law by not admitting into their deliberations the second technical analysis.

But the court went beyond the questions framed by the tannery. I cited a provision of the Environmental Protection Act requiring each authority "to consider in its decisions three elements: (1) social interest [in this case, environmental protection]; (2) existing level of contamination or hazards to the environment; (3) realistic feasibility of complying with its decision." In the court's opinion, the ministry complied with the first clause, gave inadequate consideration to the second, and failed to realistically consider the third because compliance would "necessarily lead to shutting down production at the plant, and closing the factory." According to the court, "the Ministry's task was to give due consideration to two interests in the context of the existing realities—general social interest of protecting the environment and feasibility of compliance—and to give clear and well justified preference to one of them. Since the feasibility of compliance was not taken into consideration . . . the decision was not consistent with the provisions of the law."

The ministry responded to the court's decision in September 1995 by revoking the earlier decision by the voivodship environmental authorities to close the lagoon and by directing them to reconsider the matter while "giving due consideration to two interests in the context of the existing realities—general social interest of protecting the environment and feasibility of compliance—and to give clear preference to one of them."

By the time we concluded our case study (fall of 1996) no final decision had been made, and the tannery continued using and monitoring the lagoon.

Disposal of Leather Shavings

Chromium-laden leather shavings are a byproduct of the tanning process. Over the years approximately 4,000 tons of shavings have accumulated on site in a specially designated area. Since 1980 the tannery has been required to have a permit for the disposal of leather shavings and to pay annual fees to WWOS. Sometime in the 1970s, the area originally designated for disposal of this material became full, and management expanded into an area designated as "green area" by WWOS but within the facility's perimeter. The environmental authorities continued collecting environmental fees for the disposal of shavings in this area through the early 1990s.

The trouble started in March 1994. During a routine inspection PIOS determined that disposal of the shavings in the "green area" of the premises was illegal and imposed a fine of approximately $1,000 for the 6,500 tons of illegally disposed material. The tannery promptly appealed the voivodship decision to the chief environmental protection inspector. Its main line of argument was to allege legal flaws in PIOS's reasoning and internal inconsistency in administrative actions over the years. First, the tannery argued, PIOS had no right to impose fines for material collected prior to 1980 because the enabling legislation requiring a permit for locating a disposal site—the Environmental Protection Act—only came into existence in 1980. Second, the tannery pointed out that until the 1994 inspection, no regulatory agency had ever objected to the tannery's disposal practices. Third, WWOS had collected environmental fees for the disposal for years. Furthermore, the tannery reminded PIOS that in 1986 WWOS officials had suggested moving the disposal site to the area currently questioned, on grounds that it was environmentally more suitable. Finally, the tannery objected to the classification of some of the waste as biologically active (which would affect the rate for calculating the fines).

Within months, the chief inspector of environmental protection reversed the decision of the voivodship authorities on most counts: the classification of the waste; the impropriety of imposing fines for activities that were not illegal before 1980; the tacit approval by the authorities of the tannery's activities over the years, as evidenced by collection of waste disposal fees; and the internal contradictions of the regulators' decisions, as demonstrated by earlier discussions of shifting the disposal site to the disputed location. Still, the inspector concluded that using the "green area" without an appropriate permit was illegal.

Following the decisions of the state environmental inspector, the voivodship PIOS recalculated the fines using a reduced rate on the reclassified material and counting only the waste accumulated since 1980. But the tannery, perhaps flushed by its successful legal defense of the lagoon, was in no mood for compromises, even though the amount of the new fine was trivial and even though it was moving the old waste from the "green" to the "brown" area. In July 1996 the enterprise submitted an appeal of the decision by PIOS to the State Supreme Administrative Court.

As before, the tannery's appeal strategy focused tightly on alleged administrative decisions. "Can [an agency] punish for waste storage if it also collects fees for that activity under the definition of environmental management?" asked the tannery. Citing the published official interpretation of the law, the document pointed out that fees and fines are "conceptually distinct: the purpose of fees is to create conditions conducive for good practice in using natural resources; fees are inherently derived from the idea of using natural resources in a manner [least harmful to them as] specified by the law. Fines, on the other hand, are a form of punishment for breaking the law in using the natural resources. By the act of collecting fees for storage of waste, the authorities implicitly accepted that practice as an activity consistent with the [environmental protection] law.'' Hence, imposing fines would be illogical and ''would constitute a double punishment'' on the enterprise.

At the time of this writing the case was still pending and the tannery had not paid any fines for waste storage. Meanwhile, despite these confrontations, collaborative efforts by WWOS and the tannery continued, and a mutually acceptable plan for disposal of leather shavings was found. The fresh shavings are sold to a manufacturer of gelatin, and the old shavings will be incinerated in a high-temperature cement kiln in Krakow. The approvals of WWOS offices in Radom and Krakow and a contract with the cement manufacturer have already been obtained. In interviews WWOS officials found nothing odd about helping Radom Tannery solve its difficulties even as the tannery took them to task in repeated legal appeals.

Management of Occupational Hazards

Our site visits to the tannery revealed a generally low concern among workers about mechanical accidents, chromium dermatitis and hearing loss: the operators of machinery and handlers of chromium-treated hides generally did not wear protective gloves or hearing protection devises, although both types of equipment could be seen lying about at their workstations.

The reported occupational accident rate (all types of accidents combined) at the tannery for the 1984–1995 period is unremarkable, ranging from 5 to 25 per 1,000 annually, with an average of 12 per 1,000 for the 1990–1995 period. The equivalent sector-specific national rate in 1995 was 19.5. As required by law, the facility maintains employees health records and conducts periodic health examinations. The health exams are tailored to the facility's occupational hazards: hearing test, dermatologic tests and respiratory tests for workers exposed to noise, trivalent chromium and ammonia and sulfuric acid, respectively. Based on the health surveillance, between zero and two cases of occupational diseases, including dermatitis, have been identified annually during the 1990–1996 period.

According to the director, exposure to ammonia, often in excess of occupational standards, is the most common occupational health hazard. Ammonia emissions arise from the manual preparation of solutions and from manual transfer of ammonia solutions from drums to the manufacturing equipment. The management's response is to rotate workers so that no person is exposed for more than 10 to 15 minutes at a time. The occurrence of dermatitis among workers suggests that chromium is a significant health hazard as well.

An extensive paper trail documents an almost continuous presence of regulatory authorities at the tannery during the past several years. Visits by the state labor inspectors occurred approximately twice a year. Inspections by the voivodship sanitary inspectors occurred monthly between 1993 and 1996. Several of the inspections focused on assessing the quality of the facility's testing laboratory and on ascertaining whether the firm routinely monitored workers' exposures to occupational hazards. Many post-inspection protocols note the need for improvements in lighting and noise reduction, but without making specific demands or setting clear deadlines. Other protocols require relatively small changes, such as a labor inspector's request to assign three workers to different jobs because of their apparent lack of formal qualifications for their current duties. A few protocols simply reminded the firm of its legal obligations.

We found no record of fines, either imposed or threatened, by any of the occupational regulatory authorities, even though the director readily acknowledged during our interviews that the occupational standard for ammonia was regularly exceeded. Neither did we find any comprehensive analysis of working conditions at the tannery, or any critical evaluation of the state of workers health relative to occupational standards or any strategic plan for the future. As summarized by the director, "Some things [hazards] cannot be eliminated and we must rely on interim methods [of hazards management] intended to prevent illness amount the individuals working in those hazardous conditions. The authorities understand that."

The director clearly considers the Labor Inspectorate to be a good source of technical advice and a willing responder to requests for assistance. She confirmed our initial observations, based on the written documentation, that no fines have been imposed on the facility for non-compliance with the Labor Code. When asked about her most preferred investments into the occupational health, without a moment's hesitation she named two: better ventilation and automation of the solution-making process. Neither is likely to happen, given the current financial climate at the factory. The director also acknowledged the weak safety culture among workers but had no plan to address this fundamental managerial issue.

Summary

The tannery's struggle to survive puts it in a difficult situation: to generate funds for environmental investments, it must produce more leather (between 40 and 60 tons of leather per month), but more production means more liquid waste, solid waste and air emissions. And increased waste and emissions mean more regulatory pressure. The voivodship Ecological Fund, the firm's one potential source of funds, is reluctant to invest in financially fragile enterprises. And potential private investors would find the tannery, and especially its toxic lagoon, a very risky business at best.

Many enterprises confronting such problems would give up and file for bankruptcy, but the firm has shown a remarkable will to survive, both in the free market and against environmental regulations. In doing the latter it has exhibited a highly sophisticated strategy and flawless timing. It has skillfully deployed technical, legal, procedural and social arguments as appropriate and has repeatedly turned environmental authorities' definitions, procedures and arguments back against the decisions they were intended to support. In her struggle with the administrative system, the tannery manager has demonstrated not only uncanny determination and temerity but a formidable familiarity with the laws, regulations and administrative processes. Her reading of the psychology of her adversaries, the mood of the social arbiter and the tendencies of the court has been right on the mark. She knew exactly when to be obstinate, when to portray the firm as a victim and when to remind authorities of the social consequences of the tannery's demise.

The dilemma faced by the authorities is more complex. All the parties involved in the case—WWOS, PIOS, the labor inspector, the voivodship sanitary inspector, the local SANEPID, even the mayor of Radom—know and admit that the tannery should close. But all hesitate for three reasons: first, justifying their decision in a way that will withstand the firm's shrewd and certain challenge has proven to be a difficult task; second, the authorities clearly feel a certain degree of sympathy for the feisty struggling tannery; and last, all are acutely aware of the social consequences of abolishing 430 jobs in a town already facing 20% unemployment.

The failed attempt by WWOS and the Ministry of Environmental Protection to close the tannery by revoking its wastewater disposal license shows that the environmental authorities were not well prepared (and perhaps not firmly resolved) to make a case that would withstand challenge. In his interview with us, the mayor of Radom stated that he would long ago have closed the tannery by revoking its contract for wastewater disposal with the municipal sewage works, but being an elected official, his hands were politically tied. It is an open question whether the mayor would deliver on this threat if he were in a stronger political position. Closing the tannery would make the city responsible for two environmental problems currently owned by the enterprise: the toxic lagoon and the repository of leather shavings. So long as it survives, it will be the tannery that pays for shipping the leather shavings to the cement kiln in Krakow and covers the costs of guarding and maintaining the lagoon.

In many ways the tannery is a symbol of the errors of the "old system." Built without a proper wastewater treatment plant or a minimally adequate industrial sludge disposal system, the tannery is a product of industrial policies that blatantly disregarded the environment. Now burdened by both that failed legacy and the new challenge of competing in a free market economy, the tannery and its employees are understandably viewed by some government representatives as victims.

Such sympathy is reinforced by some parties' awareness that they let things linger for too long. The city and WWOS missed past opportunities to develop a land management plan and a waste management plan appropriate for a voivodship dense with leather tanneries and thus prevented the tannery from building an environmentally acceptable sludge disposal system while public opposition was minor. This sense of collective responsibility—certainly a priceless cultural asset when compared to the horrific "blame and revenge" ethos engulfing some other former communist states—helps to explain the readiness of environmental authorities to recognize the enterprise's modest efforts to make environmental improvements and to repeatedly give it another chance.

Still another consideration is that no one agency bears the full responsibility for allowing the tannery to continue to operate. At least four different agencies—the voivodship WWOS, the voivodship PIOS, the mayor, and the ministry—have all the necessary reasons and authority to force the firm's closure. Clearly, each would prefer that the decision come from elsewhere. In this context the 1992 action by the voivodship WWOS, though unsuccessful, can be viewed as courageous. The support that came for the *voivoda* from the ministry is easier to explain because for the ministry the decision was diffuse and impersonal. In contrast the court's decision to overturn the ministry may seem surprising. But in its reluctance to aggravate social disruption and economic hardship in individual communities, the court spoke for the general mood of a country still hard-pressed by the challenges of a far-reaching political and economic transition.

CASE 3: WYSZKÓW FURNITURE MANUFACTURER FAMA

Overview

Wyszków Furniture Manufacturer Fama is located in Wyszków, a town of 21,000 people, 35 miles from Warsaw. Fama employs 630 persons. Wyszków has several larger employers, such as a glass works, an automobile manufacturer and a brewery. The factory was established in 1963 to meet the growing need for inexpensive practical furniture that was created by rapid urbanization and housing construction in Poland.

At its peak production period during the 1970s, Fama employed 1,700 people in Wyszków and another several hundred at a second facility is

Ostrow. Together, the two facilities supplied 12% of all furniture sold in Poland. Its products included all types of home furniture as well as particle board. The enterprise was known for taking various initiatives in improving working conditions and winning national competitions for innovations in occupational management. It enjoyed a good reputation among labor inspectors and sanitary inspectors.

During the 1980s demand for Fama's products fell due to the national crisis in housing construction, and the factory, burdened with debt, began to experience financial difficulties. The situation was exacerbated by the fact that management and the strong union organizations were unable to develop a mutually acceptable plan to address the declining profits. By the early 1990s the firm was approaching a collapse. During that period the environmental and occupational authorities stopped monitoring the facility because it was apparent that the company had no resources to respond to official requests.

In 1994 the facility filed for bankruptcy and was offered for sale by the state under the privatization program. At that point production ceased altogether. After an intense competition between a German investor and a partnership of three Polish businessmen, the enterprise was finally purchased by the partnership, who now comprise the management of the facility. (The successful group had astutely purchased Fama's bad debt for a tiny fraction of its face value; had the German investor outbid them, he would have become liable for the full value of the debt!) The purchase agreement included buildings, machinery and real estate but did not obligate the new owners to rehire former employees. EH&S matters did not arise in the purchase and contract negotiations.

The new owners selected a group of fewer than two hundred employees from the original workforce and resumed production in October 1994. The local authorities were friendly towards the new investors and offered relief from the tax on debt but refused other economic incentives sought by the new owners (such as reduced real estate taxes). The local bank offered a favorable loan package.

In March 1995 the factory was formally dedicated under the new ownership. Its current business plan is to develop new markets across the eastern border in Belaruss, Ukraine, and other former Soviet republics. The return on investment has so far been disappointing, but the prospects for new markets remain bright.

The enterprise employs an environmental management specialist and an occupational hygienist. The workforce is not unionized. According to the production manager, who vividly recalls the old unions' refusal to recognize fiscal realities during the last years before the bankruptcy, there is little support among the rehired employees for labor organizing. When asked about unions, the president curtly dismissed the idea.

The physical state of the plant at the time of privatization was very poor: badly deteriorated buildings, dirt, leaking roofs, obsolete manufacturing equipment, absent or dysfunctional basic amenities such as sanitary facilities, cafeterias and changing rooms. All required immediate attention and investment. These problems were noted by the local Wyszków SANEPID inspector during the first, self-initiated inspection in December 1994, shortly after production was resumed and three months before ownership was legally transferred. The SANEPID and labor inspectors would be recurrent visitors at the Fama premises.

Management of Environmental Hazards

Air quality is the primary environmental issue at Fama. Shortly after the ownership transfer, WWOS ordered the enterprise to prepare an Air Quality Assessment Document by December 1995. The document, prepared by a consultant on behalf of Fama, was submitted to WWOS in September 1995.

Using the standard procedures for such calculations, WWOS calculated the incremental impact of the plant's emissions on existing ambient concentrations of various pollutants. The total anticipated levels were then compared with the legally enforceable ambient air standards. The exercise was performed for combustion emissions from the power plant, particulate emissions from the venting systems and for emissions of 18 volatile organic compounds used in cleaning, gluing and painting operations within the factory. The analysis indicated that Fama's impact on the existing air quality in the area would lead to the violation of ambient standards for nitrogen oxides, sulfur dioxides and carbon monoxide (from the power plant), for four volatile organic solvents (from the painting operations) and for particulate matter (from the ventilation system for wood processing operations).

On behalf of the company the independent consultant proposed a plan to reduce the facility's impact on local air quality: Fama would install a scrubber on the power plant, switch to a lower sulfur fuel, combine organic emissions from the painting operation into one emission stream and increase the height of that emission point to facilitate dispersion and switch to different types of particulate filters. The proposed emission rates submitted to WWOS in the September 1995 air permit application assumed that the above technological changes would be introduced by the company in the near future.

In March 1996 the firm also presented to WWOS a two-year plan for upgrading the facility's environmental performance. The plan included a combustor for wood dust, a taller stack for air emissions, filters for air recycling and a scrubber for the power plant, all at a total cost of approximately $400,000. Both the scope and the schedule of the improvements were clearly beyond the fiscal capability of the enterprise: the plan was a useful statement of Fama's needs and strategies regarding environmental improvements but not a realistic blueprint for action. In our interviews the voivodship directors of both WWOS and PIOS stated that they never believed that Fama would be able to implement the plan in the specified period. However, the director of PIOS stated that he would defer any fines triggered by Fama's non-compliance if he was convinced that so far as its resources allowed the firm had begun to invest in the proposed improvements.

The voivodship WWOS gave Fama a three-year permit in March 1996. The emission rates and technological specifications in the permit exactly match the corresponding sections in the permit application, again suggesting that environmental authorities had confidence in the technical competence, fairness and judgment of the expert who prepared the application on Fama's behalf. The permit obligates Fama to switch to low sulfur fuel and to maintain its existing pollution control technology in good working order. The compliance dates for achieving the allowable organic compounds emission rate is the end of 1997 and for power plant and wood processing emissions, the end of 1998.

Shortly after issuing the air permit in April 1996, the voivodship PIOS conducted an inspection at the factory, including measurements of the emission rates from the power plant. The report laconically notes that the company has not switched to a low sulfur fuel and that the venting system for particulate matter has not been modernized or adequately maintained. Nevertheless, the report notes no violations of the allowable emission rates and therefore imposes no fines.

While it remains to be seen whether the company will be able to meet the deadlines set by the 1996 air permit, its ongoing burden in environmental fees is significant: the company pays $8,000 per year for the power plant emissions, an amount equal to half of the fees imposed between 1994 and March 1996 when the facility operated with a permit.

Management of Occupational Hazards

The occupational hazards at Fama include noise, fumes from painting and cleaning operations, airborne wood dust and mechanical hazards from machine operations. One of the major occupational hazards of the pre-1994 facility, formaldehyde emissions from particle board manufacturing process, is no longer an issue because the operation has been discontinued under the new ownership.

During the immediate post-privatization period in 1994, the regional Labor Inspectorate watched the facility from some distance, assessing the attitude of the new owners towards working conditions and gauging their long-term commitment to rebuilding of the enterprise. During that early grace period the authorities did not conduct formal inspections and gave the management latitude to put their house in order. Certain moves made by the management at that time, such as compensating employees fairly and on time and restoring the firm's health clinic, were interpreted by the Labor Inspectorate officials as indications of a good faith commitment to the enterprise and its employees. The president, who had previously managed another furniture factory, also enjoyed a good reputation. Altogether, the Labor Inspectorate adopted a cautiously cooperative attitude towards the enterprise.

Formal inspections of the plant began again in early 1995. According to the voivodship labor inspector, the immediate objective was to identify and promptly remove the most obvious and easily managed hazards. Numerous safety problems were identified during the first three inspections, including exposure to fumes and noise, and fines were imposed on the middle management for negligent handling for certain safety matters. The labor inspector also initiated a collaborative program with the local SANEPID (which has a testing laboratory) to monitor Fama's compliance with key occupational standards.

The local SANEPID undertook a formal inspection of the facility. The 1995 report recommended the already-initiated indoor air monitoring for airborne solvents and improvement of the physical plant, including cafeterias and workers' rest areas. The report also advised management that it needed to address the noise problem—something that would clearly require time and capital investments. The results of the indoor air monitoring program, reported and analyzed in January 1996, indicated a violation of the occupational standard for toluene: depending on the workstation, exposure levels were two to five times higher than the standard. The results of SANEPID's assessment were shared with the voivodship Labor Inspectorate, which promptly issued an order to solve the non-compliance problem, but imposed no fines. In his interview with us, the Labor Inspectorate explained, "We did not impose any fines because there was no indication of resistance to the order on the part of the firm's management."

The local SANEPID gave Fama three months to comply with the occupational toluene standard, and the enterprise responded by presenting a plan to replace toluene-based paints with water-based materials. The company also requested a one month extension of the compliance deadline so that its occupational hygienist could monitor air in the plant to assess the effectiveness of the solvent replacement. The local SANEPID inspector reported to us that she was initially inclined to grant the extension but that when an April inspection revealed that the switch had not taken place, she denied the request. During our first visit to the factory a month later, in May 1996, we found that the new technology was already in place.

The noise problem at Fama is much more difficult to solve. The relevant exposure limits are routinely exceeded by factors of two to four, depending on a location. The long-term solution, not currently considered by the owners, would require a substantial capital investment in modern machinery. When asked about their position on the issue, the SANEPID representative acknowledged that noise is a chronic occupational hazard at the plant and that the enterprise has not been in compliance with the occupational standard. However, she added quickly, "Unlike exposure to toluene vapors, these noise levels do not cause toxic effects or present a threat to life, and a permanent solution is currently beyond the fiscal means of the enterprise." For now both parties have agreed to an interim solution that relies on individual protective devices for workers, shortened workdays and job rotation.

During our interview both the director of local SANEPID and the representative of the voivodship Sanitary Inspectorate called attention to the financial fragility of Fama and its importance for the local economy. Both stressed the need to work cooperatively towards reasonable solutions. While both strongly supported the current occupational standard for noise, they were willing to allow FAMA to use interim solutions at the plant as long as the enterprise demonstrated good will and made efforts to reduce hazards to workers.

Airborne wood dust is a fire and explosion hazard at Fama but because of the relatively large size of the airborne particles, not a respiratory hazard. From the perspective of the factory management, the indoor dust problem is just one aspect of a larger modernization and energy efficiency problem. The enterprise generates a very large quantity of dust, and removing it from the indoor environment through Fama's antiquated venting system leads to tremendous heat losses in winter. These losses were partially offset by the sale of wood dust as fuel, but Fama management noted that the market for fuel wood has been declining. Since the cost of petroleum fuel for Fama's power and heating plant has been increasing, financial loses due to the enterprise's obsolete energy and ventilation systems have steadily increased. The long-term solution envisioned by Fama management is to install a new ventilation system which will remove the wood dust (for on-site combustion in an upgraded multi-fuel heat and power plant) and recycle the heated air back into the buildings. Unfortunately, Fama does not now have the capital to implement this solution.

In December 1994 the Ministry of Labor reduced the occupational wood dust standard from 4 to 2 mg/cu.m. Fama did not contest the standard per se, but joined 10 other furniture manufacturers in Poland (both state- and privately owned) in petitioning the ministry to defer its implementation by four years. The petition offered arguments on financial, technological and scientific grounds, arguing that the health hazards of wood dust do not justify the stringency of the new standard, that commonly available technology cannot achieve it and that the economic burden of the new standard would be excessive. In addition to this national-level appeal, FAMA has also appealed to the regional Labor Inspectorate to defer Fama's own compliance deadline for the new standard and presented a plan to switch to a different species of wood which is not subject to the new standard. During our interview representatives of the Labor Inspectorate said that they were favorably disposed towards Fama's petition.

Fama's participation in a multi-enterprise petition to the ministry is striking in at least three respects. First, for its modest goals: despite their many arguments, all the participants accept the validity of the new standard and ask only for a relatively brief postponement of the compliance deadline. Second, for its very small base: barely 1% of the hundreds of potentially affected wood products enterprises in Poland joined in the appeal. And third, for its rarity: although we asked about similar multi-enterprise appeals in all our interviews with corporate and regulatory officials, this was the only such effort which we were able to discover.

Our visit to the factory (with CIOP officials) confirmed the noise problem and the need to upgrade the physical plant. The general employee attitude towards safety was obviously laissez faire: workers did not wear protective clothing or shoes; hearing protection devices were available but, even in the most noisy areas, not in use; materials were stacked in aisles, creating tripping hazards; there was no sprinkler system. Nevertheless, workstations utilizing toxic volatile solvents were equipped with air monitors, confirming our earlier observation that SANEPID was seriously concerned to reduce exposure to toxic agents. The occupational hygienist who accompanied us on the factory inspection did not seem disturbed or embarrassed by the lax attitudes towards accident prevention, citing Fama's very good accident statistics. To us, the inspection suggested that the overall safety culture at Fama, both among the workers and the management, is not strong.

Summary

Fama is a struggling enterprise with promising future prospects. If it can find capital to implement the plan presented to WWOS, it will be able simultaneously to reduce the costs of energy generation and environmental fees, solve most of its environmental problems and improve occupational health and safety. But capital is also needed for improvements to the facility's manufacturing equipment and infrastructure.

Since privatization Fama has had close and frequent interactions with the occupational regulatory authorities. In our interviews it was apparent that both the regional and local authorities had a remarkably detailed knowledge of the occupational hazards within the plant. They also understood the financial fragility of Fama and its importance for the local economy. Both groups of officials stressed the need to work cooperatively towards reasonable solutions. And while they strongly supported the current occupational standard for noise, they indicated that they were willing to allow Fama to use interim solutions at the plant as long as the enterprise demonstrated good will and made efforts to reduce hazards to workers.

After a careful observation of Fama's new owners, all these authorities made a tentative judgment that management was acting in good faith and therefore adopted a cautiously cooperative mode of interaction with the firm. Both SANEPID and the labor inspector, for example, could have responded to Fama's non-compliance with occupational standards (noise) by demanding immediate compliance and imposing fines; instead, they waived fines, deferred compliance deadlines and accepted incremental improvements, such as the use of personal protective devices. Neither the occupational authorities nor Fama management seemed to have any plan to improve the weak safety culture of workers and supervisors.

The occupational authorities were clearly much less willing to compromise on hazards associated with chronic disease or threat to life. Both the labor inspector and the local SANEPID pressured Fama to quickly eliminate organic solvent exposure hazards. The local SANEPID was adept, determined and effective in pressing the enterprise to achieve this objective.

Although representatives of the local SANEPID and of the voivodship Labor Inspectorate are not legally required to cooperate or communicate, it was very clear in this case that they had shared their information and judgment extensively and often. Representatives of the two agencies were familiar not only with each others' reports and decisions but also with each others' assessment of Fama management. And both shared the view that the need for health and safety improvements must be balanced against other vital social goods.

We found that an equally close and collaborative relationship existed between the voivodship environmental authorities from WWOS and PIOS. And these authorities, too, shared the view that environmental goals must be balanced against other vital social goals. Both PIOS and WWOS accepted Fama's two year plan for environmental improvements despite their shared judgment that it was clearly unrealistic. And both suggested that, as long as Fama showed good faith by making as many of the proposed changes as its financial situation allowed, there would be no need to impose any fines.

Finally, as in the first three cases, our study of the paper trail showed that the air permit issued by WWOS was virtually a copy of the permit application. Both Fama and WWOS seem to have complete trust in the technical competence and good judgment of the ministry-certified expert who prepared the application. And the expert acted not only as a technician, but as an accepted de facto mediator between the two parties.

CASE 4: RADOM PAINT FACTORY RAFFIL

Overview

The Raffil paint factory was our second case study in the city of Radom. The firm was established in 1917 to manufacture vegetable oils and over three decades shifted towards synthetic paints and finishes. The factory is located in the center of Radom in close proximity to major residential areas. Raffil now manufactures paints and finishes, primarily for domestic industrial use. In 1996 it sold over 6,000 tons of paints, solvents and related products and reported gross revenue of $12 million. The financial situation of the facility is good, reflecting a 5–10% annual growth rate over the past three years. In 1996 the firm had 260 employees, a 13% decline from a 1990 workforce of 300.

Raffil was privatized in January 1995 as a wholly employee-owned enterprise. One of the conditions imposed on the privatization agreement by the Radom *voivoda* was an ambitious 10-year plan for environmental improvements.

The president, an engineer by training, is a 15-year veteran at Raffil. In our interviews he was clearly familiar with the environmental and occupational hazards at the facility as well as the regulatory requirements and climate in both. The environmental and occupational managers displayed a remarkable familiarity with the details of each other's jobs, projects and problems. In response to our comment on this subject, they seemed surprised by the idea that it could be otherwise. After all, they observed wryly, Raffil is a small facility with a stable and tightly knit workforce.

Despite the fact that they own 100% of the company's stock, the workers at Raffil are represented by two unions. During our site visit we observed a modern and clean enterprise without noticeable odors. We were struck, however, by the workers' seeming disregard for personal safety: slippery floors were not cleaned, too many workers rode on an open vehicle, safety glasses and hard hats were missing even in areas where they were clearly needed. The occupational accident rate in Raffil has been unaffected by either privatization or the larger changes in Polish society, consistently averaging 9.3 per 1,000 from 1987 to 1995. It should be noted that this compares very well to the 1995 national average for the chemical industry, which was 44.8 in the private sector.

Management of Environmental Hazards

Raffil is required to obtain permits for air emissions, solid waste disposal at a local landfill and groundwater use. Wastewater disposal is covered by a negotiated contract with the municipal sewage treatment facility. Solid industrial waste from Raffil (between 70 and 80 tons per year) is sent to an incinerator outside Radom. Our study of Raffil focused on the management of air emissions (the main environmental issue). The facility has two types of air contaminants: combustion emissions from its power plant and VOC emissions (7.7 tons per year) from transfer operations, building vents and fugitive emission points throughout the plant.

Our analysis of the documentary record extended backward to 1988. In that year Raffil was notified by the voivodship WWOS that ambient air standards for several volatile organic compounds were being exceeded in the vicinity of the factory (as demonstrated by recent monitoring data) and that WWOS had received many citizens' complaints about odors from Raffil. The 1988 notice instructed Raffil to undertake, before January 1991, a modernization process that would eliminate these problems.

In July 1991 Raffil notified WWOS that it had completed numerous improvements which would eliminate the odors and assure compliance with ambient air quality standards. Shortly thereafter, WWOS formally requested that Raffil submit an Air Quality Assessment Document "demonstrating the effectiveness of the measures reported by the firm" to improve the air quality in the area. The deadline for the document, which would also serve as Raffil's application for air permit, was December 1991.

In response to WWOS's request, Raffil hired a ministry-certified consultant to prepare the document. The choice of the consultant, who was recommended by the Radom WWOS, would prove to be a mistake with consequences for all the major actors in the case. Almost two years later, the consultant had not completed the document required by WWOS's order. There is no evidence that WWOS (whose director, in our interview, sorely regretted having recommended the consultant) took any punitive action in response to this flagrant violation. Indeed, in an October 1993 letter reminding Raffil that they were still waiting for the *Operat*, the authorities are quite polite. The firm's response, sent a month later, enclosed copies of letters from the consultant (who claimed unspecified "technical difficulties") and requested an extension of WWOS's deadline for the permit application. Four months later, the firm asked for another extension, again citing the consultant's failure to deliver the product and pointing to the success of its own modernization program in reducing air emissions. WWOS again responded positively, extending Raffil's deadline to the end of June 1994. But while all these polite communications were taking place between the firm, its consultant and the regulatory agency, the neighborhood was growing impatient with the facility's failure to respond to their complaints about odors. Their grassroots movement mobilized local elected officials and, in May 1994, the mayor of the City of Radom presented an official appeal to the *voivoda* to "(1) take disciplinary action against Raffil for its failure to prepare the Air Quality Assessment Document; (2) compensate neighborhood residents for exposure to noxious fumes; and (3) buy out the private properties most affected by the emissions plume from the factory."

Although we found no copy of the *voivoda's* written response to the mayor, neither the compensation, the buy-out, nor the disciplinary action took place. On the contrary, WWOS continued to show remarkable leniency towards Raffil for several more months, even protecting it from the environment enforcement authorities (PIOS). Shortly after the mayor's appeal, the voivodship PIOS imposed fines on Raffil. The facility appealed, citing both the failure of the WWOS-recommended consultant and its own efforts to make environmental improvements. Raffil's appeal was supported by WWOS, which intervened on its behalf and accepted responsibility for recommending an unreliable (albeit ministry-certified) consultant. The appeal succeeded, and PIOS suspended its fines.

The curious saga of Raffil's permit application continued for many more months. In August 1994 (more than three years after WWOS' initial request), Raffil notified WWOS that the Air Quality assessment document was almost completed and requested a September meeting to discuss the results of the consultant's analysis. But neither the assessment document nor the consultant's analysis ever materialized. Our review of the minutes of the September meeting between Raffil and WWOS revealed that the consultant *still* had not performed an air impact analysis and that the firm, with WWOS' approval, had finally decided to hire a new consultant. In December Raffil officially notified WWOS that it had selected and hired a new consultant.

Despite the patience the authorities showed Raffil while waiting for the fraudulent consultant to deliver his report, and despite WWOS's intervention to suspend fines imposed by PIOS, WWOS never waived Raffil's ordinary air emissions fees. Indeed, as shown in Table 4-4, WWOS regularly collected surcharges from Raffil for operating without a permit. For instance, the September 1994 bill for air emissions fees includes a 50% surcharge for operating without an air permit. It is not clear how WWOS could have calculated these fees without a supporting emission rate assessment, but it is clear that both the fees and the surcharges were calculated in ways which did not punish Raffil. For example, the 1994 fees were calculated only for combustion emissions from the power plant

Table 4–4.
Environmental Fees and Fines at the Raffil Paint Factory between 1990 and 1995 (in thousands of złoty)

Year	Fees			Fines
	Wastewater	Industrial and Solid Waste	Air Emissions	
1990	27.9	11.2	0.5	
1991	31.0	12.0	0.7	
1992	54.3	18.7	3.8	NONE
1993	76.4	6.3	2.2	
1994	131.7	7.3	2.5	
1995	173.2	6.3	no data	

and ignored VOC emissions from the plant. Moreover, at 50% the surcharges were only half the stipulated 100%.

In September 1995 Raffil finally submitted the long-awaited air permit application to WWOS. The new consultant had worked hard for one year: it was a carefully prepared document of comprehensive scope and high professional caliber. The application focused on six combustion emissions from the power plant and 17 VOC emissions from 79 distinct point sources in the production process, the transfer operations and the ventilation system. Applying sophisticated dispersion models to the emission rates from each of the individual emission sources and assuming that background pollutant concentrations in the neighborhood already equaled 50% of the allowable maximum ambient air concentrations and allowing for all of Raffil's technological improvements, the report concluded that ambient standards for two (of 17) volatile organic compounds (xylene and ethylene glycol) would be violated at (respectively) two and three locations in the neighboring area. Using the same assumptions, the report concluded that emissions from the power plant would exceed four of six relevant standards. The permit application recommended emission limits for each pollutant that would assure compliance with all ambient air standards. These could be achieved, according to consultant's analysis, if the firm invested in technological changes to facilitate the dispersion of pollutants and implemented a long-term production process modernization program to reduce fugitive emissions.

In November, following the submission of the Air Quality Assessment Document, WWOS inspected the Raffil plant. Two months later, in January 1996, WWOS issued a three-year air permit to Raffil. The permissible emission rates listed in the permit were identical in all details to those proposed by the consultant on behalf of the enterprise, including such details as VOC emission rates for 79 individual point sources. The permit also required Raffil to modernize its plant by the end of 1997 (without specifying the types of modernization), mandated semiannual monitoring of emissions from all pollution sources and required ambient air monitoring in the neighborhood.

Not unexpectedly, given the level of communication between the enterprise and the regulatory authorities, air monitoring for the four VOC's of greatest concern began even before the air permit was officially issued. The results exceeded allowable 30-minute limits for two compounds and the 24-hour standard for one compound, and just met the 24-hour standard for another. Since the applicable regulations allow infrequent noncompliance with both the 30-minute and 24-hour standards (but not the annual standards), whether or not the facility was violating the law was, on the basis of just one report, a discretionary judgment. The report did not find the facility in violation, but recommended continuation of the periodic air monitoring. Although Table 4–4 shows that no fines that were imposed on Raffil between 1990 and 1995 for air emissions, this was because it was operating without a permit, not because it was in compliance with air standards.

Analysis

Raffil's experience with both incompetent and competent air quality consultants neatly illuminates the professionalization of environmental assessment in Poland. The requirement that firms submit an Air Quality Assessment Document to obtain an *Operat* was first introduced in 1990. In the beginning fraudulent or incompetent consultants—like the one hired by Raffil—could operate under the umbrella of initial certification. But as both WWOS and the private sector gained experience with these consultants, the bad actors were squeezed out and replaced by highly competent professionals like the one who eventually prepared the air permit application for Raffil.

In the Raffil story we see an astounding degree of cooperation between a regulatory agency and a polluter. The appearance of an unreliable or fraudulent consulting firm on the ministerial list was not so surprising in 1990, when new procedures for preparing permit applications were suddenly implemented. But the patience exhibited by WWOS regarding Raffil's four-year delay in submitting an Air Quality Assessment Document and permit application is striking. Clearly, WWOS could not have believed the "technical difficulty" excuses repeated by the first consultant.

Raffil's patience with the situation is understandable: the lack of a permit protected it from PIOS inspections and possible non-compliance fines. WWOS's patience has more complex roots. Both WWOS officials and Raffil management pointed out to us that WWOS felt responsible for the bad advice it had given Raffil regarding the choice of a consultant. WWOS's honest admission of error is certainly admirable. Whether it justified WWOS's intervention against PIOS and local politicians to protect Raffil from penalties and community outrage is a more complex question.

Did WWOS stretch cooperation too far in the direction of co-option? Perhaps. But perhaps not. WWOS's decision to impose only a 50% surcharge on Raffil's air emission fees, when the law clearly called for a doubling in the absence of a permit, may concretely reflect its judgment that it shared responsibility for the delay. During his interview with us, Raffil's environmental director put the issue thus: "Why would they impose a fine on us if the reason that we did not have an application was that they recommended the wrong consultant? But on the other hand, we cannot really blame them either, because they relied on the ministry-approved list! Their mistake was like an occupational accident."

Another key factor helping to explain WWOS's patience with Raffil was their judgment that the facility posed no serious health or safety hazard to the neighborhood. The director of the voivodship WWOS—the same official who had attempted to close the Radom Leather Tannery in January 1992—indirectly confirmed the non-negotiable character of serious health threats by explaining his patience with Raffil thus: "I could not close the factory—there was no imminent danger to health."

CASE 5: PENCIL FACTORY ST. MAJEWSKI

Overview

Majewski Pencil Factory was esablished in 1894 in the center of Pruszków, a densely populated urban center of 40,000 located within commuting distance of Warsaw. For the next 50 years, the Majewski name was synonymous throughout Poland with high-quality pencils and other office products. Within the Pruszków community, the family-owned and -operated enterprise was a significant employer (between 300 and 400 employees) and a generous neighbor, active in the community life and in the local charities. The company's strong tradition of progressive, paternalistic capitalism was legendary, as was its employee loyalty. During World War II, the enterprise was an important center of anti-German resistance.

Majewski was nationalized in 1948, and over the next four decades little changed at the factory. The physical plant, manufacturing technology, product line and product quality survived, and Poland continued to rely on Majewski's pencils (under a new name, of course), but no capital investments were made by the state to keep the firm modern. At present more than 70% of the firm's 300 employees are women, and few of the jobs require more than minimal skills or training.

During our visit to Majewski we interviewed all the top management, including the president, the occupational manager and the environmental manager. We also met with representatives of the local union organizations, which have a strong presence at the enterprise. Our visit to the factory was a long journey back in time: the buildings, while clean and well ventilated, resembled factories from the Industrial Revolution and the heavy reliance on manual labor (even for such easily automated processes as painting, gluing and packaging) raised obvious questions about cost-effectiveness and free-market survival.

The privatization of the factory in June 1995 was unique even in the present Polish circumstances: the majority of the stock (65%) was returned to the Majewski family and the remaining 35% was distributed to the employees. But the agreement imposed heavy conditions on the new owners: for two years, wage increases were guaranteed and layoffs were forbidden.

The Majewski enterprise is in a very fragile financial state. Adverse market conditions include declining demand for pencils and growing foreign competition. The heavy burdens of the 1995 social contract with the employees have produced an almost flat salary structure, with consequent loss of the most skilled employees. Employee anxiety over the survival of the enterprise is so palpable that, in 1966, the labor union agreed to ignore the privatization contract and allow a 10% reduction in the workforce. The managing director is counting on inflation and attrition to reduce the burden of the payroll and allow him to introduce a skill- and performance-based compensation system. He frankly admits that his plan will be moot if the enterprise cannot survive the next several years.

Management of Environmental Hazards

The enterprise is required to obtain permits for its use of groundwater, its wastewater and sewage disposal its solid waste disposal and its air emissions.

Air emissions from the factory include combustion byproducts from the power plant and organic vapors from the painting operations. The VOC emissions consist of five common solvents, amounting to approximately 20 metric tons per year. The Air Quality Assessment Document and air permit application submitted to the voivodship WWOS in November 1993 contained a remarkably detailed analysis of substance-specific emissions rates and of their impacts on ambient air concentrations in the neighborhood. The analysis includes sulfur oxides, carbon monoxide, particulates, nitrogen dioxide and each individual organic solvent, from every operation and every location within the factory. Detailed calculations (using air dispersion models) are given for each chemical at four different heights above ground and at the points of maximum impact in the neighborhood.

Based on these calculations, and taking into account the background concentrations determined by the voivodship sanitary inspector (which ranged from 30% to 90% of the ambient standard), the consultant preparing the documents for the enterprise concluded that four ambient air quality standards would be violated and that full compliance would require reducing the total organic emissions by approximately 85%. According to the consultant, this could be accomplished by switching to water-based paints (not possible because of the effect of water on the wooden pencils), by installing activated carbon filters to capture the emissions or by increasing the height of the two vent stacks from 10 to 30 meters.

In February 1994 WWOS issued a permit (valid only until the end of 1994) which adjusted the recommended emission limits for some of the individual VOCs but set the combined VOC emission limit at almost exactly the level recommended by the consultant. The permit also obligated the facility to present, by the end of 1994, a plan for reducing emissions of four solvents. In the meantime, no fines were recommended. In effect WWOS gave the firm a grace period in which to find a way to reduce the emissions.

In our interviews, however, the environmental manager at the facility admitted that no steps had been taken to address the air quality issue. WWOS and PIOS documents confirmed her statement. The fees imposed by WWOS for air emissions in 1995 showed that VOC emissions had not changed appreciably from 1993 or 1994. In 1995 WWOS imposed a 50% surcharge for operating without an air permit, but no fines for noncompliance were imposed by PIOS. Our strong impression is that both sides have put the issues to rest for a few years while the enterprise struggles to survive.

When asked to comment on the relationship between the enterprise and the environmental authorities, the environmental manager first described the difficult financial state of the firm and the impossibility of considering any capital investments into environmental improvements. Then she reminded us of the relatively modest impact of the factory on the air quality in the neighborhood. "The authorities understand our situation," she concluded. "Our relationships are based on collegiality and on an implicit understanding that we must all ensure that the factory does not become an odor nuisance or a fire hazard to the neighborhood."

Management of Occupational Hazards

Accidental injuries are the principal occupational hazard at the factory. The annual accident rate during the 1993–1995 period rose from 18.2 to 19.6 to 32.0. The 1995 rate was an order of magnitude higher than the national average (approximately 3) for the private wood and wood product manufacturing sector. The firm's official report to the Labor Inspectorate blamed most of the accidents on employee negligence and failure to follow proper personal safety practices. Our tour of the manufacturing areas with CIOP officials, however, found that many of the machines lacked even the simplest safety guards.

Other occupational hazards in the facility include noise, airborne dust and organic vapors from the manual painting operations. The occupational standards for noise are exceeded at some workstations by approximately 10%, due primarily to the antiquated equipment. The labor inspector has noted the non-compliance, but no fines have been imposed on the firm. The voivodship Sanitary Inspectorate issued an order in 1995 to reach compliance by the end of March 1997, but the occupational manager we interviewed stated flatly that nothing could be done (short of equipment replacement) and that the firm had no plan to comply with the order.

The occupational hygienist, clearly concerned about the occupational health and safety of employees, complained about the dismissive attitude of workers towards personal safety and the low effectiveness of safety training. She described her cooperation with the local SANEPID efforts to monitor occupational hazards and to improve working conditions by issuing personal hearing protection devices and protective dust masks. During our tour of the manufacturing areas, we did see such devices, but very few were being used by the workers. All these initiatives came from the occupational hygienist working jointly with the local SANEPID and regional labor inspector. There was no evidence that the local unions, otherwise very active in workers' affairs, participated in the assessment of the occupational hazards or were concerned to improve health and safety conditions.

Summary

The issue of air emissions from the painting operation illuminates the regulatory climate in which the Majewski enterprise operates. The emissions were analyzed in remarkable detail, and worst case scenarios were modeled with high technical expertise. The emission rates proposed in the enterprise's permit application by its consultant were legally and technically correct but completely unrealistic given the current economic plight of Majewski. The voivodship WWOS took a critical look at the recommendations of the ministry-certified consultant and in light of its own detailed knowledge of the firm's economic circumstances, chose not to insist on the stringent emission standards that the technical analysis had produced. Instead, it adopted the strategy of allowing the firm to operate without a permit. While the lack of permit necessitated a surcharge on normal environmental fees, it also protected the factory from the much larger non-compliance fines that PIOS would have been legally required to impose on the ailing enterprise had the permit been issued as recommended. Even the amount of the surcharge reflects WWOS's sensitivity to the fragile economic state of Majewski: it was reduced by half from the customary doubling of regular fees.

This concludes our narrative presentation of the individual case studies. In the next chapter, we adopt a more analytic position and consider what lessons the cases—considered collectively—can offer.

Chapter 5

Findings from the Case Studies

Our five case studies offer a detailed view of Poland's EH&S regulatory institutions at work. This chapter presents a systematic analysis of those case studies. We offer several conclusions—few definite, many unexpected and several in the form of hypotheses for further work. But first, a cautionary note about the limitations embodied in our cases.

CASE SAMPLE LIMITATIONS

Although our case studies focus on precisely the kind of medium-sized enterprises that constitute the core of Poland's growing private economy, it is important to note some ways in which these firms do not represent the whole of that economy.

First, the firms that we studied all have between 200 and 1,000 employees. None is the sort of very small enterprise that might, at least temporarily, escape the notice of local or regional EH&S officials. And none is an EH&S dinosaur of the sort that might, like the fading steel industry, attract the personal attention of central authorities in Warsaw. Rather, these are the kind of firms whose EH&S performance is the daily professional concern of local and regional EH&S officials.

Second, the firms that we have studied are all located within 200 kilometers of Warsaw. Despite the fact that they are in four different voivodships (and so enabled us to study and compare four different sets of regional authorities), none of these firms is located in the relatively pristine Northeast or the heavily polluted Southwest. It is possible that local or regional EH&S officials in those regions might approach the balancing of EH&S goals against other societal goals with a different perspective than the many officials we interviewed.

Third, the firms that we have studied were all privatized in or before 1995. They are among the many formerly state-owned enterprises that were included in the first large wave of state divestiture. It is possible that, the new private economy now having been so successfully launched, EH&S officials might be less inclined to offer "breathing room" (in the form, e.g., of reduced fines and negotiated compliance schedules) to new or newly privatized enterprises in the future.

Despite these three limitations, our five case studies involved a wide range of business arrangements, market and capital circumstances, production technologies, EH&S problems, local social roles and regulatory contexts. They offer a detailed and informative look at EH&S regulation in Poland's emerging economy.

INFORMATION RICHNESS

In the course of our interviews with key EH&S actors, we were repeatedly surprised by the richness—both in scope and in detail—of their knowledge of virtually all aspects of the cases we were investigating. Senior and mid-level enforcement and inspection officials were consistently able to discuss in fine detail not only the material dimensions of the cases (e.g., the firm's particular technologies, occupational health and safety issues, environmental hazards and performance and all the firm's activities in these areas for at least the past several years) but their economic, social and personal dimensions as well. WWOS and PIOS officials were intimately familiar with each enterprise's ownership structure, market and capital situation, past and present role in both its municipality and its surrounding neighborhood, and corporate philosophy. They could discuss in significant detail the outlook, strengths and capabilities of key corporate managers, including EH&S personnel; and in the few cases where those individuals were relatively new to their positions, they were able as well to describe the reputation which the managers had built at other firms. The occupational health and safety officials whom we interviewed seemed no less well informed. In addition, the regulatory paper trail which we studied was dense, technically expert, detailed and typically comprehensive, even by OECD standards.

Reluctantly, we came to believe that the remark of one labor inspector, made during one of our first rounds of interviews, was no idle boast: "We know the firms in our district like the backs of our hands."

The roots of this information richness, at least on the side of the EH&S inspection and enforcement officials, are not mysterious. One root is found in the professional and institutional continuities described in the next section of this chapter. A second, which became apparent in both our study of the paper trail and our interviews with corporate and regulatory personnel, is this: the facilities that we studied were subjected to on-site visits by either sanitary or labor inspectors or by PIOS personnel with remarkable frequency. For example, the Radom Tannery had been visited 34 times in the three years preceding our first visit. This regular, routine presence of EH&S regulators in production facilities, a legacy of Poland's socialist period, is possible because of the tight network of cooperating institutions. It is one of several such legacies which (as discussed in Chapter 1) were deliberately preserved—against many foreign experts' calls for total institutional restructuring—through the incremental reform of Poland's EH&S regulatory system. A third root (also discussed below) is the informal but very efficient way in which officials attached to different EH&S agencies share information. As another labor inspector keenly observed, "None of our regulatory and enforcement instruments would work if we did not have such a well-developed network."

This "network" is the human instantiation of another dimension of information richness: namely, the extent to which key EH&S actors know *each other*. In all of the four voivodships in which we worked, it was clear that the senior and mid-level environmental officials at the licensing and enforcement agencies knew each other well. They understood each other's institutional mission and regulatory philosophy. They were aware of each other's activities at the firms under study and had clearly shared their judgments about the extent to which the firm in question was (or was not) acting in "good faith" to make EH&S improvements. Representatives of the occupational health and safety agencies seemed to be no less familiar with one another. This personal dimension of "information richness" is clearly rooted in the continuities discussed below.

The EH&S enforcement and inspection agencies and the corporate EH&S professionals we interviewed also displayed a broad and detailed familiarity with the economic, social and personal dimensions of EH&S management. All were able to price past and proposed EH&S investments from memory. All knew how much their firms were paying (and had paid in recent years) for fees, fines and surcharges. All had a detailed knowledge of the ways in which EH&S regulation had evolved since 1989. All were familiar with prevailing regulatory standards and their likely evolution in the future. All had a realistic sense of their firms' relative importance to their communities as an employer.

Finally, the corporate EH&S professionals whom we interviewed were also remarkably familiar with their local and regional enforcement and inspection officials. They had detailed (and, in our judgment, impressively accurate) opinions regarding the officials' attitudes toward their firms and toward EH&S regulation in general. We were particularly struck by the fact that both the corporate managers and EH&S professionals gave no evidence of being surprised by the decisions that regulatory officials had taken regarding their firms. Even in the midst of a dramatic societal transition and of incremental regulatory reform, the firms had sufficient knowledge of their government counterparts to predict EH&S regulatory decisions with considerable accuracy.

CONTINUITY THROUGH TRANSITION

Our case studies revealed that, despite Poland's rapid and simultaneous political and economic transition, there has been a remarkable degree of continuity among key EH&S actors. This continuity is found in institutional structures and relationships and in both regulatory and corporate personnel.

Since 1989 EH&S enforcement agencies have grown incrementally, especially at the junior and staff levels. The senior and mid-level administrators and technical experts who today staff local and regional EH&S agencies, however, are essentially the same men and women who staffed the same agencies before 1989. All of the senior staff whom we interviewed—in four different voivodships—were long-term professional employees, typically with at least 10 to 15 years experience with their current institution.

Working relationships between the several different EH&S institutions have also continued essentially unchanged through Poland's transition. Thus, WWOS officials continue to license enterprises and to calculate and collect environmental fees, all the while depending on PIOS to verify reported emissions data. As a senior WWOS official commented during one of our interviews, "There is no legally mandated relationship between WWOS and PIOS, but here at the voivodship level, mutual communication and cooperation are just facts of life."

We found similarly close communication and cooperation between the agencies in the occupational health and safety domain. Although significantly more numerous (and independent of political control) than before 1989, the labor inspectors continue to work closely with both the regional sanitary inspectors and the local SANEPIDs. The Labor Inspectorate officials whom we interviewed took respectful note of their reliance on both local SANEPID personnel (for technical laboratory services and detailed information about developments in particular factories) and the Sanitary Inspectorate (for occupational disease prevention activities).

These continuing patterns of cooperation and communication, however, include a continuing, and familiar, "disconnect" between the occupational agencies and the environmental agencies. In Poland, like in other OECD economies, the agencies charged with environmental protection are neither required nor expected to communicate or cooperate with the agencies charged with occupational health and safety, although we did see small signs that it may begin to change. At the national level, CIOP—Poland's premier scientific institute for occupational health and safety research and standard-setting—has recently launched some educational initiatives for corporate professionals which will deal with environmental as well as occupational issues.

Basic institutional structures have also continued during the first post-1989 decade, with incremental adjustments. The most notable exception involved increasing the independence of the enforcement agencies. On the occupational side, the State Labor Inspectorate was removed from the control by the Ministry of Labor (see Figure 3–2), while on the environmental side the voivodship environmental inspectors were made to report directly to the chief state environmental inspector rather than, as before 1989, to their *voivodas* (see Figure 2–1). The institutional reorganization in effect since January 1999, discussed in Chapters 2 and 3 for environmental and occupational sectors, respectively, opens this continuity to a test. On the one hand, the reversal of the reporting context of the voivodship environmental inspectors back to their *voivodas* (Figure 2–5) and the shift in the lines of authority over SANEPIDs and voivodship sanitary inspectors to elected county managers (*starostas*) and *voivodas*, respectively (Figure 3–5) will probably have relatively minor implications for the daily functions of these institutions and their abilities to make independent decisions. This is partly because the appeals of administrative decisions continue to be protected from the *voivodas'* and *starostas'* influences, and because of the highly technical nature of the work of these agencies. Furthermore, the recent reorganization requires no major personnel shifts. On the other hand, the elimination of WWOS offices and decentralization of the facility environmental licensing decisions to county level may have a major effect on the continuity of personnel and the decision-making process, especially regarding small and medium-size facilities that dominate a particular local economy.

We found that the level of managerial and professional continuity in the five enterprises was almost as high as in the government agencies. The presidents of Radom Tannery and Raffil are long-term employees of their firms. The presidents of Drumet and Fama are relatively more recent arrivals to their firms (having taken their positions in 1989 and 1994, respectively), but each came from a senior post in a similar enterprise. Only the president of Majewski Pencil Factory was a real newcomer, having been a high-level government administrator before joining the firm. But even he was recruited to Majewski because of deeper cultural continuities: he is a member of the obviously still revered family which owned the firm until its nationalization after World War II.

Continuity is even clearer among the firms' EH&S professionals. Without exception, the occupational and environmental managers we interviewed had been with their firms long before the 1989 transition. The majority were already in their current position by that time. Even at Fama, where privatization was preceded by bankruptcy and the laying off of all the firms' employees, the new owners rehired both the environmental and occupational specialists into their previous positions.

At each of the five firms the main environmental and occupational issues have also continued from before 1989. Neither the larger social and political transformation in Poland nor the specific changes associated with privatization have, by themselves, solved old EH&S problems or uncovered new ones. At Drumet the problems of hydrochloride emissions and occupational lead exposures were decades old, and the ambitious technological innovations that would eventually solve them—although clearly accelerated by concerted regulatory pressure—were already under way by the mid 1980s. The occupational accident rate at Drumet did begin declining around 1989, but as noted in our case narrative, this seems to have been the result of initiatives by the new management team. At the Radom Leather Tannery small gains have been made in solid waste disposal and chromium recovery from wastewater, but the most serious environmental problems—wastewater disposal, sludge disposal and the toxic lagoon—are the same today as before 1989. Odors and air emissions have been a problem at Raffil since it began to manufacture paints; although significant progress (probably related to increased regulatory pressures) has been made in reducing VOC emissions, the issue of odors continues to disturb the firm's neighbors. At Fama environmental and occupations performance has been improved by reducing VOC emissions and exposure, but the inadequate ventilation system and the dirty, inefficient power plant continue to plague both the regulators and the new management.

The continuity enjoyed by the bureaucracy and firms during the 1989–1999 period is likely to decline in the future. As discussed earlier, the 1999 reorganizations of the environmental and occupational bureaucracies, especially with regard to facility licensing, will lead, at the minimum, to personnel changes and, possibly, to some changes in the decision-making processes. In addition, both the size and professional mobility of the entrepreneurial and environmental professional classes are increasing in response to sustained economic growth in Poland and the maturation of the environmental protection system. For example, the chief voivodship inspector for Warsaw voivodship told us during a spring 1999 interview of a 50% personnel turnover at his institution! While not surprising, the process of replacing the personnel retained from the communist era by those coming of age in the new political and economic realities may have unanticipated future consequences for the EH&S system in Poland. We return to this topic in Chapter 7.

REFORM WITHIN CONTINUITY

One might expect that so much human and institutional continuity would be a powerful barrier to EH&S change, and that it would create a massive bureaucratic inertia with a systematic bias toward preserving the pre-1989 status quo—including lax enforcement practices, disregard for official standards and so forth. Indeed, as we noted in Chapter 1, this is precisely what numerous authorities in the West predicted (with reason) in the early years after 1989. Nevertheless, *this is not what we discovered.* Especially in the environmental protection area, regulatory officials seem to have embraced reform and quietly persuaded corporate actors that the old days are gone forever.

Numerous incremental reforms—including especially the 1991 standardization of the licensing procedures for industrial enterprises—have been successfully implemented and have clearly increased the confidence of voivodship WWOS and PIOS officials. Our case studies suggest that these officials are committed to enforcing the law and making the new system work. All four of the voivodship WWOS offices in our study have required submission of Air Quality Assessment Documents and permit applications (*Operats*) from the firms in their jurisdictions and have issued environmental permits promptly after receiving the *Operats*. All are collecting environmental fees and imposing surcharges on enterprises operating without licenses.

As our cases illustrate, none of these reforms was implemented blindly. The bureaucratic *homo sovieticus* predicted by some experts appears nowhere in our case narratives. Implementation (as discussed below) has been sensitive to the particular circumstances of each case. PIOS and WWOS are serious about their missions. The decision by the Radom WWOS to force the closure of the beleaguered tannery, despite the difficult social context, illustrates its determination to stop practices which present unacceptable environmental hazards. The ongoing air monitoring around Drumet and Raffil illustrate PIOS's seriousness about strengthening enforcement practices. PIOS officials were quite candid about these changes in their interviews. One remarked bluntly, "In the past, an inspector could defer decisions indefinitely. That has changed!" A second PIOS official put it this way: "The fundamental difference between then and now is that if we do not do our job properly, we will be fired."

In the occupational health and safety sector the reform and implementation picture is less impressive. Clearly, there have been gains. All the labor inspectors we interviewed pointed out ways in which their institution had been strengthened since 1989: increased enforcement powers, increased funding, higher job performance expectations and increased morale were mentioned repeatedly in our interviews. One labor inspector told us bluntly, "I have had to completely change my work ethic since 1989. I am not the same man I used to be!" In the context of the interview, there was no reason to doubt that he was telling the truth—or that he was proud of the change.

The occupational health and safety personnel we interviewed clearly believe in strict enforcement of occupational standards when the risks of occupational disease or serious accidents are high. For instance, the understaffed and underfunded SANEPID in Wyszków, where staff lack even a photocopier and must rely on manual analytic techniques in their outdated laboratory, did not hesitate to assert itself against Fama management regarding workers' exposure to toxic fumes. After an initial period in which it monitored exposure levels at individual workstations, the SANEPID successfully pressured the firm's new management to phase the toxic organic solvents out of the manufacturing process.

But despite these gains, our interviews and site visits suggest a deep difference between the occupational and environmental reform processes. In the environmental arena, both managers and regulators clearly expect significant performance improvements. In the occupational arena expectations seem to be fundamentally unchanged in a number of key performance areas, including accident rates, worker safety practices and compliance with many occupational standards. Indeed, in our interviews both managers and regulators seemed to share a fundamental pessimism about the possibility of substantial progress in these areas. Our site inspections at Fama, Raffil, Majewski and the Radom Tannery provided abundant evidence of the problem.

Personal protective devices were ubiquitous at all the facilities we visited but were rarely in use. Even at job stations with obvious mechanical, toxic or noise hazards, workers actually using goggles, masks, respirators or hearing protectors were far outnumbered by workers simply carrying the devices on their persons. The same observations applied to mechanical safety systems. At Fama and Majewski, for example, we observed saw guards and press fences lying next to the equipment from which they had removed. In some respects Drumet seems a notable exception to this pattern: nearly every worker at a hazardous workstation used the appropriate safety devices and nearly all were wearing steel-toed shoes. But even Drumet's accident rate is very high by U.S. standards. Moreover, the Drumet story confirms our general impression of worker safety culture: safety improvements were imposed on the workers by top management with little input from the labor inspector or the Committee for Occupational Safety and Hygiene, and *against* workers' and unions' complaints about loss of compensation benefits. Low employee participation in intra-facility occupational safety organizations was the norm at all sites.

In part the roots of these health and safety problem are found in social factors. The communist regime's practice of mandating compensation bonuses for hazardous working conditions no doubt tended to select and preserve a weak industrial safety culture. In part the roots of the problem are legal and institutional. The full enforcement of Poland's very comprehensive and stringent occupational standards would require (and at Drumet, *did* require) capital investments on a scale which most firms simply cannot contemplate. Fama's situation with regard to the dust and noise standards is illustrative here. It is also clear that the occupational regulators lag behind their environmental counterparts in developing the institutional experience and mechanisms that would enable them to negotiate explicit incremental agreements with firms regarding the implementation of the Labor Code.

Faced with all these difficulties, occupational health and safety officials now seem to engage in a kind of triage. Code infractions which represent clear and serious threats to human health and safety provoke strict and swift enforcement. For less serious infractions, inspectors may choose to delay the imposition of fines, to push facilities toward incremental improvements or temporary solutions (such as personal protective devices or job site rotation) or to overlook the infractions entirely. In our interviews, managers and regulators offer remarkably similar descriptions of the current modus vivendi. The director of the Radom Leather Tannery put it this way: "Some hazards cannot be eliminated, and we must rely on interim methods to prevent illness among the individuals working in those hazardous conditions. The authorities understand this." For his part one government official said, "If non-compliance with an occupational standard presents a threat to human health and life, we stop the operations immediately. Otherwise, we sometimes overlook delays, . . . give enterprises a chance to comply, . . . and seek reasonable solutions." The striking contrast between approximately monthly inspections of the tannery by occupational personnel and the nationwide shortage of labor inspectors (as discussed in Chapter 3, less that 5% of all firms registered with Labor Inspectorate are visited in any given year) is consistent with these views.

Thus, while environmental authorities are both attacking the worst problems (as at Radom Tannery) and driving other firms at the leading edge to new performance levels (as at Drumet), occupational authorities seem still to be entirely focused on raising the bottom. We would conclude that in both their activities and their institutional culture the environmental agencies have so far been more successful in implementing post-1989 reforms than have the occupational agencies.

NEW PARTICIPANTS: TECHNICAL CONSULTANTS

The codification and implementation of procedures for obtaining air and water pollution discharge permits since 1991 has introduced an important new class of actors into the environmental regulatory arena: technical consultants specializing in the preparation of environmental impact documents and permit applications (*Operats*). These legal and procedural changes created a major business opportunity for the technical community, and several dozen individuals and firms have already been certified by the Ministry of Environment to enter the field. At the same time early corporate and government experience with these new consultants is leading to the elimination of incompetent or unscrupulous actors and increased professionalization. The completed *Operats* that we reviewed were carefully prepared and meticulously detailed. The exceptional experience of Raffil, which hired a fraudulent consultant in 1991, illustrates the professionalization process: the consultant has been removed from the approved list and drummed out of the business.

Our cases reveal that *Operat* consultants exercise considerable influence on the conditions of pollution permits for small and medium-size companies. There seem to be several reasons for this. First, by law, the *Operat* must not only assess the environmental impact of a firm's emissions but must also recommend both specific emission rates and specific means of meeting ambient standards. By its nature an *Operat* compels the technical consultant to move beyond the mere presentation of quantitative data. Second, an *Operat* can have a major economic impact on a firm: it defines necessary technological changes (and hence, capital demands), provides the basis for calculating environmental fees and sets the benchmark for potential fines. Third, and perhaps most important, the preparers of *Operats* seem to have become trusted de facto mediators between the authorities and enterprises.

Our case histories revealed that neither enterprises nor regulatory officials are likely to question the analyses and recommendations of the consultants. But neither were they ever surprised by those recommendations: all the parties clearly knew what the others doing. In three of the four instances in which we were able to compare the *Operats* to the resulting emission licenses (Drumet, Raffil and Fama), the licenses were identical to the *Operats* in both content and format. In the fourth case (Majewski), regulatory officials made minor changes to the air permit—confirming that they had indeed reviewed the *Operat* closely but again, trusted largely in the consultant. In all cases, the permits were issued within days of the *Operat's* submission.

The Drumet air permit case illustrates the influence that these new technical consultants can have on regulatory decisions in the context of the newly standardized application procedures. The emission rates demanded by WWOS between 1986 and 1990 were not technically justified, were repeatedly challenged by Drumet and led to years of inconclusive confrontation. But the 1993 permit, issued on the basis of a consultant-prepared *Operat*, was accepted by all parties despite the fact that (imposing stricter limits than Drumet had sought but allowing higher emission rates than WWOS had favored) it was completely pleasing to none.

The remarkable role played by the ministry-sanctioned consulting industry is, so far as we are aware, unparalleled in any other Western EH&S regulatory system. In Poland the effect has been to add another participant to the EH&S decision-making process for small and medium-sized enterprises—and not inconsequentially, one who is structurally invested in preserving the cooperative mode of decision making which we discuss below.

MISSING PARTICIPANTS: THE TRADE UNIONS

One of the striking findings of our field work was the conspicuous absence of organized labor from efforts to set or reach EH&S goals. With regard to occupational health and safety, local unions were frequently the key obstacle to progress. In case after case local unions were much more concerned to perpetuate compensation bonuses for hazardous working conditions than to militate for, or even cooperate in, the elimination of the hazardous conditions themselves. At best the local unions were simply silent on occupational health and safety. In no case were they pressing to put health and safety issues on management's agenda.

The role of organized labor regarding environmental matters has traditionally been indirect, and one might have expected that it would be no different in the Polish context. But in fact, Solidarity made environmental issues a central theme in its tenacious and decisive campaign to overthrow the communist regime, and put environmental issues at the top of the national agenda during the 1989 roundtable discussions. Given that dramatic history, labor's rapid and apparently complete retreat from concern with environmental issues—bluntly confirmed by national labor leaders during our interviews in Warsaw—is more than striking. Nevertheless, it seems clear that in the context of Poland's emerging free-market economy, labor unions have decided that high visibility on the national political stage and advocacy for social benefits, employment security and fair wages are their main priorities. For the present the trade unions seem to regard investments in EH&S protection as direct threats to their social agenda.

Whether this posture is tenable in the longer term—or what its effects will be if maintained—is not clear. We are aware of no other large industrialized free-market democracy in which the EH&S protection system has been notably successful without the support (let alone despite the opposition) of organized labor. How the Polish EH&S system will be able, over the longer term, to bring labor back to the table (or in the worst case, to deal with its continued indifference) remains to be seen.

INDUSTRIAL ATTITUDES: OWNERSHIP BUT SCRUTINY

Our interviews with enterprise managers and their senior staff provided an opportunity to explore their attitudes toward the EH&S regulatory system in general. Quite consistently, and at least somewhat surprisingly, the managers at all five firms were distinctly accepting of the current system of environmental fees. This is not because the fees are trivial; at Fama and the Radom Leather Tannery the fees represent a significant fraction of corporate revenue. Nor is it because the managers had a universally accepting attitude toward government-imposed business costs: the same Fama manager complained bitterly about the economic burden of government decisions and about the failure of local authorities to reduce his tax obligations. Drumet managers also offered us a long list of grievances about WWOS decisions, including unreasonable emission rates, inflexible compliance schedules and unfair consideration of high background pollutant levels when calculating Drumet's permissible emissions. Nevertheless, when asked if they thought that the system of environmental fees was fair, they answered, "Of course."

Despite their grumbling about the way ambient pollution levels caused by neighboring enterprises is used in the calculation of their permissible emission limits, enterprise managers also regard the process of setting emission limits as fundamentally fair and justified. When we asked the environmental managers at both Drumet and Raffil whether government authorities should ignore background pollution levels when setting emission rates, they answered "no" and immediately cited the public health effects of such a change as a justification. When we asked the same individuals whether they thought that the current ambient air pollution standards were too stringent, they again said no.

This general acceptance of the system does not mean that firms will not challenge individual decisions which they regard as unfair or unduly burdensome. As described in Chapter 4, the Radom Leather Tannery used sophisticated legal and procedural arguments to overturn an entire series of decisions by WWOS and PIOS. Drumet enjoyed less success, but nevertheless challenged WWOS repeatedly. And both Drumet's and the Radom Tannery's negotiations with municipal authorities regarding sewer contracts displayed a sophisticated determination to win favorable terms. Nor are these cases exceptional: in our interview with the deputy minister of environmental protection, he complained bitterly about the flood of appeals—as many as 200 in one day—which he had received from industrial enterprises regarding WWOS and PIOS decisions. And firms are increasingly willing to challenge administrative decisions by hiring specialized legal and technical services (see Chapter 3).

Nevertheless, despite their willingness to challenge individual decisions made within the EH&S protection system, the corporate managers whom we interviewed were clearly invested in the system's success.

Conspicuously absent from the corporate-regulatory interactions which we studied was the kind of large scale, organized lobbying activity so commonly found in the U.S. system of environmental standard setting. High level officials within the ministry described to us a sophisticated lobbying campaign mounted by key firms in the energy generation sector in an effort to influence emission standards and another seeking to divert some of the funds collected from environmental fees and fines to local activities. But the only evidence that we found of any similar activity by smaller firms was FAMA's petition to the Ministry of Labor (with a small group of similar furniture manufacturers) requesting a delay in the implementation of a new occupational standard for one type of wood dust.

NEGOTIATION VERSUS CONFRONTATION

In the next section of this chapter we discuss the manner in which EH&S officials endeavor to balance multiple competing goals. In this section we describe the most salient feature of their regulatory style and practice: namely, a very clear and decisive preference for negotiated solutions rather than command-driven confrontations.

Without exception every single regulatory official whom we interviewed stressed the belief that negotiation and cooperation were the preferred modes of operation and that confrontation was the enforcement tool of last resort. Over and over, in interview after interview, we heard statements such as: "If we see good faith on the part of the management, we make an effort to prevent their financial collapse," or "A repressive regulatory system does not work," or "A punitive system is not effective. We need to work with enterprises over a long periods of time to make incremental improvements. Confrontation does not solve anything."

The cooperation between firms and labor inspectors is especially notable because it is in one sense *pre*scribed and in another sense *pro*scribed by the Labor Code. According to the code, a labor inspector has two basic responsibilities: to enforce the law, imposing fines and other sanctions as stipulated; and to provide advice and technical assistance to firms. All the occupational managers whom we interviewed spoke appreciatively about the amount and quality of the advice and technical assistance they received from the inspectorate, and about the willingness of the inspectors to help them solve technical problems in order to comply with the Labor Code. One company manager remarked: "Whenever I have a problem, the Labor Inspectorate never refuses to help. I truly appreciate their cooperative attitude. I can call them any time, day and night, no matter how small the problem." None of the labor inspectors whom we interviewed saw any conflict in these dual roles or admitted to any worry about performing them. But it is at least possible that these apparently—albeit not necessarily—conflicting roles might undermine the inspectors' effectiveness as enforcers and so contribute to the relatively slower pace of reform in the occupational health and safety arena.

The systematic preference of EH&S regulators for cooperation and negotiation has its risks, as we saw in the Raffil case, where the authorities' reluctance to hold Raffil responsible for the bad actions of its *Operat* consultant led to many years' delay in the production of the required *Operat*. It also has its burdens. In particular, it requires that regulators make very well-informed decisions about whether a firm's management is (or is not) cooperating and negotiating in "good faith." The importance of showing good faith was emphasized repeatedly by the officials we interviewed and was echoed in the remarks quoted just above. Enterprise managers and EH&S professionals were emphatic in their remarks on this subject, repeatedly voicing the opinion that they would continue to enjoy a cooperative relationship with the authorities *only* so long as they made good faith efforts to implement EH&S improvements as quickly as their firms' circumstances allowed.

BALANCING MULTIPLE VALUES

Our case studies gave us a unique opportunity to examine the ways in which regulatory institutions and their personnel actually balance environmental and occupational objectives against other important societal goals.

It is clear from all our data that the single most important factor governing EH&S authorities' willingness to negotiate and cooperate is their assessment of the attitude of the industrial managers. And such assessments require a rich empirical basis. Thus the occupational authorities in Wyszków carefully investigated the reputation of Fama's new president and then watched his actions carefully before adopting a cooperative posture towards working conditions at the furniture factory. Fortunately for Fama, the new management passed the "good faith" test. In his interview with us the labor inspector described his subsequent actions this way: "We did not impose any fines because there was no indication of resistance by management." The occupational health manager at one of the firms we studied described the general situation this way: "Labor inspectors are not policemen but rather a resource. However, *that continues only as long as we show a positive attitude and make improvements*. If we do not complete agreed upon tasks by the designated deadlines, we will be fined. The [regulatory] institutions have had to change during the past few years; they cannot be as lenient as they used to be. Two years ago we were fined for not having periodic medical exams for all our workers. That would *never* have happened before!"

In all our cases and interviews, occupational health and safety officials displayed a keen sensitivity to the financial and cultural difficulties experienced by some enterprises in meeting the strict norms laid down by the Labor Code. Despite their minimal *formal* authority to balance the enforcement of these norms against other considerations, it is clear that they regularly do engage in precisely such balancing and utilize a variety of informal tools to implement their decisions, including partial or delayed enforcement, as discussed in the preceding section. As one labor inspector put it in our interview, "Closing a factory or a manufacturing operation is a very difficult decision requiring consideration of multiple factors. After all, closing a facility means punishing many innocent parties—employees, suppliers and customers!" The Labor Code does not provide for any explicit negotiation of operating conditions or terms of compliance; by implication, its punitive stipulations seem flatly to proscribe such negotiations. Nevertheless, the labor inspectors we interviewed do calibrate their enforcement efforts to the firm's circumstances, the management's level of good faith and their own assessment of the level of risk to workers' lives and health. Thus, in the cases that we investigated, the paper trail included numerous recorded violations of various provisions of the Labor Code but very few penalties or fines.

By comparison to the occupational protection authorities, the environmental authorities have much more discretionary leeway and many more opportunities to balance competing objectives. Our case studies indicate that they use their freedom and opportunities extensively. Although the Environmental Protection Act did not (as of 1999) specifically instruct regulatory agencies to consider implementation costs or technological feasibility when setting performance standards or issuing licenses, it did include the provision that all regulatory decisions should "consider [both] social interests and the realistic feasibility of compliance." (See the excerpts from the court's interpretation of this provision in our description of the Radom Leather Tannery in Chapter 4.) Moreover, the institutional design of the environmental regulatory system—which separates the licensing and enforcement processes—provides regulators more opportunities to seek balanced and negotiated solutions. It also gives them some very persuasive tools. Small changes in the terms of an enterprise's environmental permit, for example, can turn modest environmental fees into much more burdensome fines or replace fines with much smaller fees.

In particularly difficult cases authorities may delay administrative decisions and allow a facility (such as Majewski) to operate without a license, providing a window of opportunity in which especially hard tradeoffs can be considered. Although nominally required to impose a 100% environmental fee surcharge on firms operating without an *Operat*, officials have the additional discretionary power (which they used in two of our five cases) to reduce those surcharges as well. Finally, PIOS has formal authority to conditionally extend a firm's implementation schedule based on its assessment of the individual case circumstances.

We found that that environmental authorities, like their occupational counterparts, were keenly sensitive to the economic and social impacts of their decisions and used the full range of their formal and informal balancing mechanisms to take account of these impacts. Based on our interviews and case studies, it appears that environmental authorities shape their judgment to serve three principles: First, they will push firms to achieve the strictest level of pollution control that they judge to be technologically and financially feasible; second, they will prevent acute threats to public health or environmental quality regardless of the impact on a firm; and third, they will reward firms demonstrating a good-faith commitment to environmental improvements with more flexible treatment. The chronology of Drumet's air pollution permit shows the first two principles at work. Over time, the Wrocław WWOS made more and more stringent demands for hydrochloride emission controls because it judged the enterprise to be financially and technologically able to meet the new standard and because it was very concerned about the public health impacts of the high hydrochloride levels in the neighborhood around the plant. Not being able to regulate the fugitive hydrochloride emissions, it addressed the second issue by imposing extremely strict limits on emissions from identified point sources.

Majewski and Fama presented a different of situation. Neither firm was financially able to implement the pollution controls recommended by their respective *Operat* consultants, but their emissions did not present a significant public health or environmental hazard, and management (especially at Fama) had demonstrated a good-faith resolve to make improvements. Accordingly, in both cases the authorities adopted a much less stringent enforcement approach. Fama received its air permit, despite the fact that WWOS knew the consultant's proposals for technological modernization could not possibly be implemented in the time frame specified, and PIOS cooperated by not imposing any fines when the first milestones were not achieved. Meanwhile, both PIOS and WWOS continued to monitor Fama closely. In the Majewski case WWOS saw a firm on verge of bankruptcy and quietly allowed the existing permit to expire without demanding a new one. Then the authorities halved the mandated surcharges, increasing Majewski's environmental fees by only 50%.

Raffil was perceived to be and treated as an intermediary case. The firm is financially healthy but has limited capital resources. Because WWOS authorities accepted partial responsibility for having recommended a fraudulent *Operat* consultant, Raffil was originally given a very long grace period; now that the *Operat* has been processed, it is apparently being treated "by the book." The new *Operat* obliges the firm to make significant investments in pollution control improvements, and the extensive air monitoring begun immediately after the permit was issued suggests that PIOS intends to enforce the agreement without further delay.

The Radom Tannery obviously presented the most difficult balancing problem. On the one hand, the voivodship WWOS understood the enterprise's financial peril and the importance of 400 manufacturing jobs an in economically depressed area. At the same time, WWOS judged the facility—and especially the toxic lagoon—to be a major environmental threat. The problem was made even more complicated by two additional factors. First, the firm's financial and environmental problems were legacies of the discredited past, fundamentally beyond the control of present management. Second, WWOS's only viable "management" option was to close the lagoon: but that would lead directly to the facility's demise, to painful layoffs and to a future in which *no one* would be responsible for maintaining the lagoon. WWOS's long delay in taking action is testimony to the complexity of the decision. For years WWOS accepted the firm's efforts to improve its pollution control technology as signs of good faith, even though all the involved parties understood that the small improvements did not significantly reduce the adverse environmental impact of the enterprise. Still, WWOS deferred punitive actions and gave the enterprise an extended grace period in which to attempt an economic recovery that would provide capital for substantial environmental investments. Ultimately, nearly all prospects of such a recovery having faded, the magnitude of the environmental threat overbalanced the broader social considerations and WWOS attempted to stop the tannery's operations. It is a staggering irony that WWOS's decision was ultimately overturned by a judge who found that it had given insufficient weight to the very same social considerations which had for so long delayed its decision!

To sum up, our cases studies reveal that environmental authorities (and to a somewhat lesser extent, occupational authorities) routinely seek enforcement strategies which balance EH&S goals against other vital societal objectives. Their balancing decisions give special weight to an enterprise's demonstration of a good-faith commitment to EH&S improvements and are informed by a remarkably detailed knowledge of the technical, economic and social circumstances of each case. Despite the lack of any formal mandate to do so, officials at different agencies routinely share information and coordinate their activities. So long as all parties demonstrate good faith, interactions between corporate officers, regulatory professionals and independent *Operat* consultants are characterized by high levels of trust, knowledge, communication and cooperation. Key corporate and regulatory actors have an intimate knowledge of each other's agendas, capabilities and EH&S philosophy. That knowledge—which is part of what we call "information richness" and without which the whole complex process of informal and formal balancing of multiple societal objectives would collapse—is the result of many years of professional interactions. It has been preserved through Poland's political and economic transition by the many continuities discussed above. It is an accumulated capital which the architects of Poland's regulatory reform have wisely protected.

CONCLUSIONS

In Chapters 2 and 3 we offered a description of the occupational and environmental protection systems in Poland which was based on an analysis of laws, policies and institutions as well as on interviews with key policy makers within and outside of government. That analysis discovered many similarities between the evolutionary paths of the environmental and occupational protection systems. In both cases there had been substantial legal and institutional continuity through the early years of societal transformation. In both cases legal and institutional reforms had been incremental.

The post-1989 reform of the EH&S protection system, we noted, was guided by four crucial assumptions: first, that the reasons for the system's past failure were well understood and essentially external to the system itself; second, that the core elements of the laws, policies and institutions inherited from the past were fundamentally sound and still well adapted to Polish society; third, that the powerful disincentives which undermined the system's performance under the communist regime had disappeared with that regime; and fourth, that given the disappearance of those external factors, modest reforms in the legal and institutional domains would enable to system to improve its performance dramatically.

Our initial analysis also found two important differences between the environmental and occupational protection systems. First, in the environmental arena, all key actors acknowledged that the system had failed monumentally under the communist regime. In the occupational arena, no similar sense of failure was apparent. There were notable reports that large numbers of workers had been exposed to occupational hazards exceeding legal limits (Academy of Social Sciences 1987; Frąckiewicz 1989; Silesian Scientific Institute 1989), but the stringency of Polish exposure standards (compared to European and U.S. standards) made these data difficult to interpret. And accident rates, which seemed to be only slightly higher in Poland, are similar to other OECD countries. As a result, while the need for change in the environmental arena was universally acknowledged, the need for reform in the occupational arena was rarely mentioned. Second, even under the communist regime, environmental protection had been a topic of academic and policy debate, so the system (and its failings) had been subjected to intensive analysis; no such tradition of analysis or debate, inside or outside the government, attended the occupational protection system. Consequently, post-1989 reforms in the occupational arena have not been driven or guided by any broad consensus; they have been almost exclusively initiated and designed by policy leaders within the central government administration.

Our initial analysis, grounded in laws and policy documents, could not tell us whether the assumptions which had guided regulatory reform were sound or whether the post-1989 reforms based on them had effectively changed the system's performance. Neither could that analysis tell us much about the ways in which the EH&S protection system was embedded in Polish society or about the extent to which it was well- or ill-matched to the values, beliefs and traditions of the individuals and groups who make, implement and enforce EH&S decisions.

Our case studies, however, permit us to do three things which our initial analysis could not: (1) to test the observations made in our preliminary analysis; (2) to explore the embeddedness of the EH&S protection system in its larger social context; and (3) to assess how well the reformed EH&S institutions and policies serve their intended objectives in specific situations.

The cases confirm and extend our earlier findings regarding the significant continuity of the emergent EH&S systems with the past—not only in the legal and institutional domains but in the professional, personal and social domains. These continuities provide critical support for key parties' efforts to find non-confrontational, cooperative solutions to EH&S decision problems.

As our discussion above indicates, however, these continuities are being challenged from two directions: by the 1999 political and administrative redistricting of Poland and by the generational changes among the entrepreneurial and environmental professional classes. The cases also reveal that the key participants in EH&S protection—corporate managers, regulatory officials and independent consultants—are technically skilled and politically astute. They share ownership of the system but do not hesitate to confront each other over individual decisions within the system. The environmental and occupational authorities whom we studied were committed to implementing the post-1989 reforms and accomplishing their respective institutional missions; they were able to implement EH&S policy in a way which balanced EH&S goals against other vital societal considerations within the context of a cooperative, information rich and very case-specific decision-making process. Local and regional authorities utilize these strategies far more often than would be expected on the basis of an analysis of official policies.

The cases provide considerable support for the four key assumptions underlying EH&S reforms. Especially in the environmental arena, careful incremental reforms clearly *have* been followed by significant changes in the behavior of key EH&S actors. Regulators, firms and the increasingly influential technical consultants all seem to have adapted their behavior to fit the new policies. Several factors seem to be involved in this successful adaptation. First, the administrative mission of each regulatory agency is clearly articulated and well understood. Second, administrative professionals support that mission and are committed to achieving it. Third, industrial managers and professionals accept the reformed EH&S system (including the system of environmental standards, the facility licensing process and environmental fees and fines) as fundamentally legitimate. Fourth, voivodship environmental authorities (and to a lesser extent, local and voivodship occupational authorities) use an array of formal and informal procedures to balance the implementation of EH&S goals against other societal concerns and do so on the basis of a detailed knowledge of each individual facility. Fifth, the implementation of EH&S reforms has been distinctly assisted by the emergence of a cadre of independent technical experts—*Operat* consultants—who are able to serve as informal mediators between firms and regulators.

Clearly, there is a notable discrepancy between the relatively inflexibility of official laws and policies (as discussed in Chapters 2 and 3) and the broad and sophisticated discretion exercised in the field by local and regional environmental authorities.

Nationally, the environmental protection system seems to operate on two levels. Large and environmentally burdensome firms such as steelworks and electrical generating facilities attract the attention of the Ministry of Environment and are subject to transparent and explicitly codified regulatory decision making. Small and medium-size firms which do not present major environmental burdens are handled by local and regional authorities (with the assistance of technical experts), and the decision-making process is adapted to the specific circumstances of the case. (It should be noted that proposed changes at the national level would open the way for the ministry to engage in more flexible negotiated decision making—so-called "implementation plans"—as well.)

Finally, in the several ways delineated in the earlier sections of this chapter, the cases confirm that the successful reform and performance of the Polish environmental protection system—at least with regard to its regulation of small and medium-sized firms at the voivodship level—is vitally connected to the larger social context in which it is embedded. In the occupational protection arena, the role of the social context is equally important but not equally supportive. This is partly due to the role of labor unions and partly due to the obvious fact that unless worker safety culture can be improved, the system will never succeed.

In the occupational arena, unlike the environmental arena, *changes in the social domain have not kept pace with reforms in the legal and institutional domain*. The new Labor Code has reduced the paternalistic role of the state in an effort to share responsibility for occupational safety with workers, worker organizations and employers; however, our case studies suggest that the first two of these groups have not yet responded at all and that only a minority of the third have responded well. A deeply entrenched poor safety culture, apathy on the part of workers' organizations and inactivity among the majority of employers have combined to leave the task of improving occupational protection in the hands of regulators. Confronted with the task of implementing an extremely stringent set of standards in this social context, and given little formal authority to engage in negotiation, savvy and committed government regulators have thus far accomplished more than one might reasonably expect but far less than successful reform will ultimately require.

characteristics go far in explaining why the system in Poland is operational.

Nevertheless, the sharp discrepancy between the common (and reasonable) pessimistic hypothesis and our own research findings requires further verification of the data on which these finding are based. This is especially important since our observations are based on the experience of only a small number of firms and may not be representative of industry as a whole. In addition, the case studies and policy analyses are more revealing of the attitudes and practices of regulators than of the firms.

This chapter describes the design and the results of a survey conducted in order to confirm and extend for a wider range of firms and industries in Poland the initial findings of the policy analyses and case studies.

RESEARCH DESIGN

A questionnaire survey was mailed in 1997 to a random sample of privatized manufacturing firms. It excluded state-owned manufacturing facilities. While many state-owned firms are, in practice, the heaviest polluters, the majority are economically uncompetitive, technologically backward and destined for closure. Our focus is on the private economy that already constitutes approximately 60% of industrial output and represents the future industrial base of the country. The State Ministry of Privatization in Poland maintains a complete registry of privatized firms. From this population we were able to obtain a strictly random sample of 300 manufacturing firms in 47 (of 49 total) administrative districts (voivodships). Previous research has demonstrated that the regulatory process often operates differently for large and small firms (see, for example, Brehm and Hamilton 1996). Small firms, for example, often receive less attention than the large firms, even though collectively small firms often are responsible for substantial air, water and waste emissions. For this reason we chose to stratify the sample into three size categories: fewer than 21–100 employees, 101–250 employees, and more than 250 employees.

The questionnaire survey was written in Polish and was approximately eight pages in length. It was pre-tested on a small number of manufacturing firms. Standard survey procedures were used to maximize the response to the survey. From the 300 firms finally contacted, we received a total of 109 completed questionnaires, for a response rate of 36.3%. There were no statistically significant differences in response rates by size of firm. The responses represented 37 voivodships. The districts not represented by the responses were all among the least industrialized in Poland.

The mean size of the firms in the sample survey was 257 employees.

On average, 22.4% of output was exported to the European Union and 4.9% to other foreign markets; the remaining 62.7% of output was sold within Poland. Larger firms, with more than 250 employees, had a significantly higher level of exports (a total of 41.1% of output was exported). The export orientation of manufacturing firms is important because the environmental standards of the European Union are thought to be a significant driver of improvements in environmental performance in Poland. We also hypothesized that the relative importance of a firm to its local economy will influence the attitude of regulatory authorities. Where a firm is the largest and dominant employer in a community (*gmina*), regulatory authorities are likely to come under greater pressure to consider employment impacts in making decisions. Of the 109 firms in the survey sample, 9.3% were the largest employer in their *gmina*, 36.4% were one of the 10 largest employers, and the remaining 54.3% were one of many employers in their *gmina*.

In general, the survey questionnaire sought to document the operation of the EH&S regulatory system in Poland. Detailed information was obtained about the licensing process, whether firms were in compliance with environmental standards, the response of authorities to firms that were out of compliance, the pattern of fees and fines paid by firms, and the role of key participants (e.g., regulatory officials, unions, employers) in the regulatory process as well as the firms' overall views on the regulatory system. Our focus was upon the regulatory process, rather than upon EH&S performance and environmental quality per se. While we did not collect data on actual emissions, we did document whether firms were in compliance with EH&S standards. Where firms were out of compliance and were required by regulators to change operating procedures, we documented whether the necessary changes were being implemented.

Because the survey respondents were a random sample of manufacturing firms drawn from all sectors of industrial production and all parts of the country, the level and type of air, water and waste emissions varied widely among participating firms. Some firms were engaged in manufacturing processes that were energy, materials and waste intensive (e.g., leather tanning and metal finishing); other participating firms had more modest environmental impacts but significant health and safety concerns (such as furniture assembly). Figure 6–1 shows the percentage of participating firms that had emissions sufficient to require permits for air emissions (54.1% of firms), industrial wastewater (30.3% of firms) and solid waste (43.7% of firms). All firms were subject to regulation for noise, occupational injuries and other worker health and safety concerns. Among the sample firms, noise was cited by 60 (55.0%) firms as their biggest health and safety problem. Exposure to chemicals, dust and other

Figure 6–1.
Percentage of Firms Requiring Permits

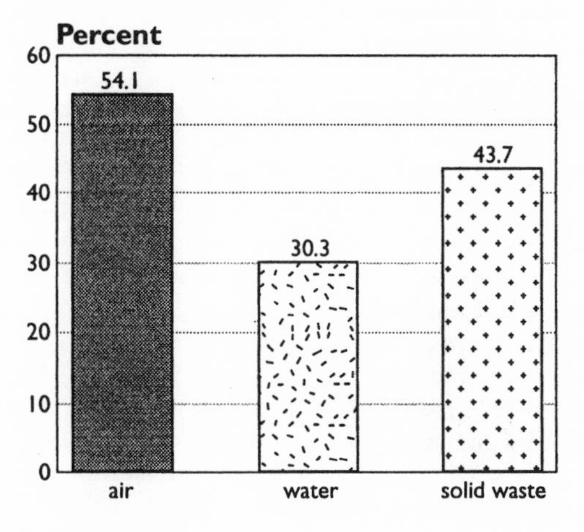

hazardous substances was cited by 32 (29.3%) firms, followed by accidental injuries (13.7% of firms).

As we might expect, large firms were more likely to have emissions that required permits than small firms. In the case of air emissions, for example, 80.0% of firms with more than 250 employees required air permits, as compared to only 26.5% of firms with 100 or fewer employees (difference significant at the 0.01 level of confidence).

RESEARCH FINDINGS

We begin with the operation of the licensing system. Effective monitoring and assessment of air, water and waste emissions is the bedrock of an effective regulatory system. As written in the law, the licensing system in Poland is a challenging, rigid and comprehensive process that holds manufacturing firms to detailed emissions and occupational health and safety standards set by central government. The EH&S standards are tough and the process of assessment is resource intensive.

The Permitting Process

Our first test is whether firms actually participate in the permitting process. Would firms actually comply with the permitting process, simply operate without permits (as often occurred prior to 1989) or seek to overturn the permitting process in the courts, parliament or other venues? The survey data reveal that with few exceptions firms agree to the detailed inspections and technical assessment that are the basis for the

Figure 6–2.
Percentage of Firms Failing to Meet Standards

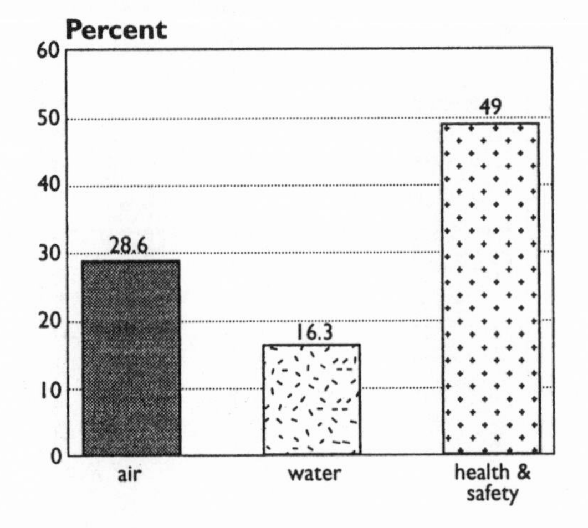

permitting process. In the case of air emissions, for example, a total of 63 of the sample firms completed the technical assessment (*Operat*) required for obtaining an air permit. This total includes *all* of the 59 firms that had air emissions subject to regulation. The remaining firms did not have air emissions that required a permit. Among the participating firms, only four reported that in the past five years they had operated without the legally required air permits. This result holds for both large and small firms.

Note that completion of the licensing process does not in itself imply that firms meet all the relevant EH&S standards. Figure 6–2 shows the percentage of firms whose air, water and waste emissions exceed allowable standards. Of the 63 firms completing the *Operat*, 18 (28.6%) reported that the operation of their factory would result in emissions that exceeded allowable air quality standards. Fifteen of 92 (16.3%) firms reported that their wastewater emissions violated standards set by municipal sewage works, though this percentage was much higher among larger firms with more than 250 employees (emissions from 30.0% of large firms violated standards). And 51 of 104 (49.0%) firms reported that occupational health and safety standards (e.g., for noise) were not met at their firm. Note that health and safety problems were found to be particularly common among large firms. Among the sample firms with more than 100 employees, 65.7% reported that their factories did not meet health and safety standards. The high rates of non-compliance with occupational health and safety standards reflects the toughness of the standards in this area and the lack of formal policy instruments for

Table 6–1.
Changes Required to Meet Emission Standards

	AIR	WATER	HEALTH/SAFETY
Small changes in operating procedures	11	4	13
Minor investment in equipment/materials	17	4	30
Major changes or major investment	7	7	20
Standards could never be met	1	1	7

stepwise enforcement as well as a weak safety culture among many workers and employers.

Table 6–1 provides information on the scale of investment and changes in operating procedure required to bring firms into compliance with EH&S standards. In many cases, minor changes in operating procedures or small investments in technology would bring firms into compliance with air, water and waste standards. But major changes would be required by seven of the sample firms to meet air quality standards, by seven sample firms to meet wastewater standards and by 20 firms to meet occupational health and safety standards. Several firms indicated that it would never be possible for them to meet current EH&S standards. Among small firms, or firms that are only marginally profitable or worse, even modest investment in technology or equipment is a major expense.

And yet, despite the tough EH&S standards and the sometimes major investments required to meet these standards, we find broad acceptance of the licensing process among the sample firms. When asked in the questionnaire survey as to their overall views on the regulatory process, only 17 of 107 (15.9%) firms stated that they found existing environmental regulations and regulatory procedures too strict. The remaining firms reported that these regulations and procedures were about correct (68.2% of firms) or too lenient (15.9%). Similarly, while 26 of 107 (24.2%) firms reported that occupational health and safety standards were too strict, the remaining majority reported that standards were about correct.

Thus our survey data indicate that the process of EH&S permitting is operational and is widely regarded as legitimate by the firms involved.

In the context of a recent history of widespread disregard for EH&S standards and procedures, this is no small achievement. At the same time many firms fail to meet existing EH&S standards. The pattern of broad acceptance of EH&S standards and of the rigorous permitting process is consistent with the findings of our case studies. In those case studies firms identified both the professional and flexible response of regulators to EH&S concerns as well the technical competence and independence of the environmental assessment process as important factors in their acceptance of the regulatory process.

The Response to Non-compliance

Our second test concerns the response of firms and of regulators to non-compliance with EH&S standards. As indicated above, a high percentage of firms fail to meet one or another EH&S standard. Would regulators enforce standards and secure the required improvements in EH&S performance? Would firms make the necessary investments to achieve these standards? In statutory terms the enforcement branch has modest flexibility in responding to a case of non-compliance, especially in the area of occupational health and safety. What we find in practice is a set of varied and flexible responses that are apparently closely tied to the financial and technical capabilities of the firms involved.

In the questionnaire survey we asked firms to report on the response of authorities to non-compliance with EH&S standards. Table 6–2 shows the results for non-compliance with air emissions and occupational health and safety standards. Note that the participating firms were invited to select all responses that applied. In the case of non-compliance with both air and health and safety standards, a minority of firms were allowed to continue operation with no changes. One firm was closed by regulatory authorities. Most firms were required to make specific changes or submit a plan with a specific timetable as to when required changes would be made. Thus the dominant response is to secure a timetable for achieving compliance and to look for evidence of progress towards this goal in the form of investments and actual changes in operating procedure. Note that we found no significant difference across size of firm or relative importance within the community in the pattern of response by regulatory authorities. These results are fully consistent with the findings of our case studies. In the case studies we found a striking pattern of regulatory authorities searching for ways to push environmental improvements within the limits of the financial and technological resources of individual firms. Regulators pressed hard where a firm could reasonably be expected to meet and even exceed EH&S standards. If firms showed evidence of good-faith efforts to meet stan-

Table 6–2.
Response of Regulators to Non-compliance

	AIR	HEALTH/SAFETY
Allowed to operate with no changes	8	7
Specific changes required to be made	17	36
Worked with us to find a solution	3	4
Imposed deadline to make changes	14	38
Required closure of offending process	0	1
Imposed fines	6	6
Required a plan to solve the problem	8	22
Ordered the factory closed	0	0

dards, they allowed firms an extended timetable within which to meet these commitments.

The imposition of fees and fines was also part of the response of regulators to non-compliance with EH&S standards. This policy tool was also used by municipal sewage works for firms failing to meet wastewater standards. For most firms, however, fees and fines remain a modest part of operating costs. In the case of air emissions, for example, of the six firms receiving fines, the fine totaled on average just 0.19% (i.e., less than 1%) of total operating costs. Among all the sample firms, environmental fees were on average 0.5% of total operating costs. We conclude that in most cases fees and fines are not the primary instrument promoting compliance with EH&S standards.

The format of the questionnaire survey did not allow for the detailed technical analysis required to establish the actual effects on pollution and health and worker safety of these enforcement activities by EH&S regulators. But firms out of compliance with EH&S standards were asked to record whether they had implemented the changes required by regulatory authorities. Of 31 firms responding to this question, 19 (61.2%) reported that they had already made the requisite changes, 10 (32.2%) reported that they were working on making the changes. The remaining two (6.4%) firms challenged the finding of the regulators in the courts. One firm appealed to regional authorities on the basis that meeting the requirements would create economic hardship and substantial unemployment. A second firm argued that the EH&S statutes had been incorrectly applied in their case. Both appeals were successful and reflect adept strategy on the part of the firms involved, a feature that was also observed in our earlier case studies (Brown, Angel and Derr 1998). In the remaining cases, surveyed firms responded that they had already made the required changes in technology and operating procedures or were currently implementing the changes.

Thus regulators have responded to widespread non-compliance with far greater flexibility and sensitivity to local technical and economic circumstance than is provided for in the law. While we are not in a position to assess fully the results of this flexible, firm-specific approach, our case study research indicates that both regulators and firms view the approach as a sensible and successful strategy for securing improvements in EH&S performance within existing socioeconomic constraints. Regulators display a high level of sensitivity to the financial and technical capabilities of firms, pushing hard where firms have the capability to meet high standards and showing flexibility where financial and technological resources are limited (see also Brown, Angel and Derr 1998).

Information Richness

The tendency towards case-specific and flexible decision making is predicated upon the ability of regulators to maintain detailed information on the operation of factories. Absent such information, regulators have little basis on which to make decisions as to the economic and technical feasibility of improving EH&S performance or the likelihood that firms will act in good faith and follow through on plans submitted for bringing facilities into compliance with EH&S standards. And yet the burden of monitoring facilities and maintaining records of compliance with multiple EH&S standards is potentially very large. In the case of Poland we find that regulators at the local and regional levels have a high level of familiarity and active presence in the factories for which they are responsible. It is this familiarity with the factories at the *local* level that is the basis for making flexible judgments on how far and how fast to push firms in improving EH&S performance.

In the questionnaire survey we gathered information on the frequency with which firms were inspected by different regulatory organizations in the previous year. The local health inspectorate (SANEPID) visited on average 2.5 times, the Labor Inspectorate 1.5 times and the environmental enforcement branch visited on average 0.6 times. Inspections by environmental enforcement regulators were more frequent at large firms with more than 250 employees (an average of 0.9 visits per annum), as compared to firms with fewer than 100 employees (0.15 visits). These differences are statistically significant at the 0.01 level of confidence. Note that while SANEPID has responsibility for health, it is also involved in calculating allowable air emissions by establishing the background levels in the vicinity of a facility. Indeed, a high degree of interaction among regulatory organizations is a feature of EH&S protection in Poland. Thus on average the factories in our sample were visited approximately four times a year by one or other branch of the EH&S regulatory system. By way of comparison, McGarity and Shapiro (1993) indicate that only one in 25 workplaces in the United States has *ever* been visited by an OSHA inspector. The national statistics from the Labor Inspectorate validate the findings of the survey. Thus the Inspectorate reports that in 1997 an average annual inspection rate of manufacturing firms employing more than 20 persons was 1.6 (State Labor Inspectorate 1998), compared with our inspection rate of 1.5.

In the survey we also inquired about the fraction of firms registered with the Labor Inspectorate that are inspected annually. Here, the national statistics also converge with the survey results (within a factor of two). Thus, according to the 1997 national statistics for small, medium and large manufacturing firms, the fractions visited were: 36.8%, 52% and 60.1%, respectively. For all sizes combined the number is 40.3% (State Labor Inspectorate 1998). The corresponding numbers derived from the survey are: 62.5%, 60%, 80% and 67.2%. The difference between the survey and the national statistics is very likely due to a response bias of the firms that chose to participate in the survey. The participating firms are likely to be more aware of, and more active in, their environmental and occupational health matters, possibly due to a greater regulatory scrutiny. In general, the considerable convergence between the national statistics and the survey results gives support to the validity of the methodology and results of the survey.

Discussions with individual firms confirm that in Poland regulatory authorities at local level are very familiar with the operation of factories the extent of any EH&S problems and the scope for achieving performance improvement within a given time frame. The general pattern is that of a regulatory system that maintains a frequent presence in factories and has a reputation for providing assistance as well as for pushing for improvements in EH&S performance.

Regulatory Responsibility

One of the striking features of the EH&S regulatory system in Poland during the period of communist rule was the heavy dependence upon the state as the primary driver of improved EH&S performance. This was especially the case in the area of occupational health protection, where the state both mandated very specific standards and procedures and was the primary advocate for worker health and safety. In conjunc-

Figure 6–3.
Groups Active in Health and Safety Protection

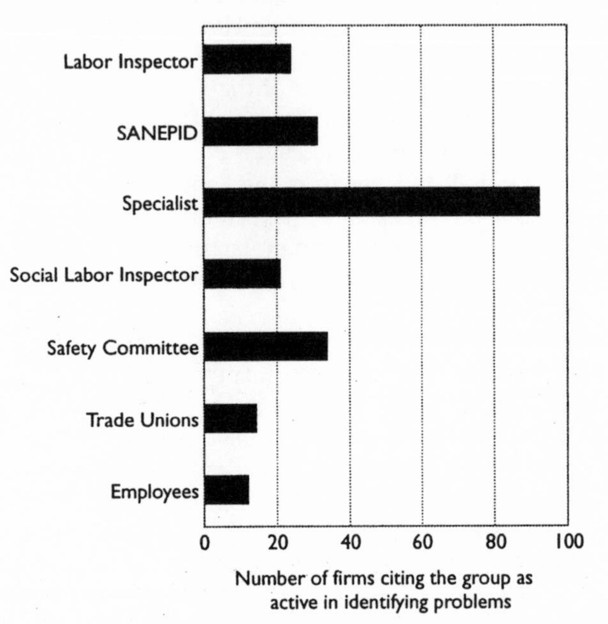

tion with the process of privatization, the state sought to shift some of the responsibility onto firms and onto the workers themselves. This was in some ways a risky strategy in that it depended on the willingness and capability of firms, workers, unions and other groups to assume this responsibility. At the same time the extensive presence of health and labor inspectors in factories was an important feature of the regulatory system.

In order to explore this aspect of regulatory reform, firms were asked in the questionnaire survey to identify which two groups in addition to management were very active in identifying health and safety problems at the factory. Figure 6–3 shows the results for the sample firms. The two most frequently cited sources were occupational health and safety specialists employed by the firm (cited 92 firms) and the firm's safety and hygiene committee (cited by 34 firms), followed by state labor inspectors and local health authorities (cited by 24 and 31 firms, respectively). By contrast, both individual employees and trade unions are rarely cited as active in identifying health and safety problems. These results suggest that regulatory authorities and occupational safety professionals continue in their long established roles as the principal advocates of workers' health and safety. While the recently established Committees for Safety and Hygiene show signs of activity, by and large the workers and their representative organizations have been slow to assume responsibility for working conditions.

Our interviews with regulators and union leaders suggest that unions are currently more concerned with issues of unemployment and compensation and are not strong advocates for enhanced health and safety protection (Brown, Angel and Derr 1998). It is not uncommon for organized labor to oppose investments in occupational health protection, with the latter seen as competing with their primary concerns of employment security, compensation and social benefits. This position is evident, for example, in the strenuous opposition of unions to the elimination of hazard pay (additional compensation for working under harmful conditions). In our survey, 68 of 109 firms (62.4%) reported that they currently have a system of hazard pay. Fifty of 104 (48.0%) reported that they thought the system of hazard pay should continue. With regulatory reform of the Labor Code shifting greater responsibility for occupational health and safety to firms themselves, it is unclear where the main drivers for continued performance improvement will come from.

CONCLUSIONS

The survey results are consistent with the findings of policy analysis and case studies. Specifically, the survey data indicate that the firms overwhelmingly participate in the process of environmental permitting by applying for and obtaining operating permits. The environmental and occupational policies and procedures are also widely regarded as legitimate by the firms involved.

The survey results also confirm our earlier observations that the regulatory authorities are quite familiar with the firms under their jurisdiction. The survey participants reported frequent inspections of their premises, especially by occupational inspectors. The data suggest that the authorities often seek non-confrontational, case-sensitive implementation, as we suggested earlier in this research work. This flexibility was especially evident when we investigated the cases of environmental noncompliance. In these cases the authorities demonstrated the willingness to negotiate stepwise implementation plans and extended deadlines. On the other hand, there appears to be the capacity for asserting the legal authority by imposing fines and by ordering facility closure, as was the case with one respondent. Additionally, the finding that the majority of reports of occupational non-compliance involved the standard for noise is consistent with our earlier observation that the authorities demonstrate selective tolerance for different types of occupational hazards. We would expect much lower frequency of non-compliance in cases of life threatening or incapacitating hazards.

The survey results also indicate that challenges to the emergent regulatory system remain. First, as we have seen, many firms are not in compliance with exposure standards, especially in the workplaces. The survey results are consistent with the views expressed by the firms and regulators in case analyses that firms are often out of compliance simply for lack of the financial resources to make the requisite investments in new process technologies and protection and abatement equipment. We also attribute this state of affairs to the toughness of standards and in the occupational arena to the lack of flexibility in implementation and enforcement.

Second, as noted in the preceding chapters, the retreat of the state from its dominant role in occupational protection has indeed left a void in the collective sense of responsibility. While the recently created Committees for Occupational Safety and Hygiene are showing modest signs of activity, the profiles of both organized labor and workers remain very low. Finally, the survey data suggests that environmental fees and fines are generally too low to be effective market-based policy instruments. This finding did not surface in our case studies, interviews with policy makers or analysis of statistical data. It is, however, consistent with observations made by other authors (Sleszyński 1998; Bluffstone and Larson 1997; Toman et al. 1994). We suggest that further studies are needed to clarify the role of environmental fees and fines in pollution abatement in Poland.

Implicitly, the flexible approach that regulators have taken toward firms that are out of compliance is predicated on a premise that economic competitiveness and EH&S protection within the privatized economy are correlated. Over time, economically successful private firms will invest in EH&S protection and improve EH&S performance; economically unsuccessful firms with poor EH&S records will eventually go out of business. In this way economic growth and EH&S protection are seen as complementary, rather than mutually exclusive, goals. To the extent that the private economy continues to grow and that prosperous firms do indeed improve their EH&S performance, this strategy represents something of a "win-win" for Poland. But the strategy remains untested under conditions of economic recession, and the level of future commitment of private firms to continuous EH&S improvements remains to be determined.

Chapter 7

Synthesis

One of the lessons to be drawn from the past three decades of EH&S regulatory experience around the world is that societies expect more of an environmental regulatory system than protection of the environment, more of an occupational health and safety system than protection of workers and the community. To be sure, a successful EH&S system must meet societal expectations for environmental and occupational protection; but typically it must do so in ways that support other societal goals.

Thus it is no coincidence that criticism of EH&S regulatory policy within the OECD is just as likely to be about the unintended economic and social consequences of regulation, such as constraining impacts on technological innovation or the failure to take account of local economic circumstances, as about the effectiveness of environmental and occupational protection measures. And within many developing economies weak enforcement of EH&S regulation often reflects fears that improvements in environmental performance will only occur at the cost of reduced economic competitiveness and static socioeconomic welfare. The ideal regulatory approach for many policy makers would yield improvements in both environmental and socioeconomic performance, and it is the tantalizing prospect of such "win-win" opportunities that dominates much of the public discourse on regulation and on "greening of industry." As a baseline, we have proposed this definition of success: an EH&S system succeeds if it advances EH&S objectives without imposing unreasonable social and economic cost, and does so in ways that enhance rather than undermine the pursuit of other societal goals.

In general it is fair to say that much existing regulatory practice is not calibrated to this optimal performance goal. Whether by legal mandate, institutional culture or custom, regulatory institutions typically externalize these concerns to the political and social domains in which societal goals and priorities are set. Similarly, there have been few systematic assessments of the effectiveness of regulatory practice structured explicitly around such balancing of competing objectives (see, for example, Davies and Mazurek 1998). In this book we have sought to identify the characteristics of regulatory systems that enhance the capacity to pursue improvements in EH&S performance in the context of multiple, sometimes competing, goals. Six key dimensions of regulatory structure were highlighted for analysis: clarity of performance expectation; appropriateness of policy instruments; information richness; case-sensitive implementation; the capacity to learn and profit from change; and broad "ownership" of the regulatory systems by different societal groups (business, communities, policy elites and the regulators themselves). We also stressed the likely importance of social context and the fit between regulatory approach and the capacities and characteristics of individual societies for the success of EH&S systems. While many other authors have cited the importance of socioeconomic context, our work offers initial explanations of why the match between regulatory approach and societal context is so important to regulatory success.

We chose to examine these broad ideas empirically through a detailed case study of EH&S regulatory reform in Poland over the 10-year period following the collapse of communist rule in 1989. Poland was selected for study because it provided a strong and focused test for the capacity of an EH&S system to yield improvement in both environmental and health and safety performance and socioeconomic welfare. The country inherited from the period of communist rule some of the worst environmental pollution in Europe, especially in the heavily industrialized region of Katowice-Krakow, and pressure to improve environmental conditions was an important part of the reform movement of the 1980s. At the same time, the commitment to privatization and economic reform fueled public demands for improved socioeconomic welfare and for expanded political and economic opportunity. As a consequence, the reform program undertaken in 1989 explicitly acknowledged that while economic growth and environmental and occupational protection could not be a zero-sum game in Poland, neither would it always be a "win-win" game (Jendrośka 1998; Ministry of Environmental Protection 1991; Bolan and Bochniarz 1994; Karaczuń 1997; Novak 1996; Sachs 1993). The policy reformers were charged with the development of mechanisms for accommodating the two mutually desirable national goals, even at the expense of making short term tradeoffs. Because public pressure and policy debate framed the challenge of regulatory reform in Poland in these terms, the case provided an important opportunity to examine those aspects of regulatory practice that facilitate the achievement of improved EH&S protection in the context of other societal goals.

At a time of fiscal austerity and competing political and social priorities on the national agenda, this developmental philosophy presented the architects of EH&S regulatory reforms with a formidable challenge. Our study shows a considerable level of success in achieving the EH&S objectives. Numerous policy and organizational changes were implemented without significant resistance or social conflict. In most cases, firms prepare environmental impact assessments and submit applications for permits, authorities issue permits in a timely manner and collect fees for pollution and resource use, firms make an effort to comply with regulatory requirements and, if necessary, administrative courts arbitrate conflicts between firms and the regulators. The administrative authorities appear to be invigorated by the reforms and demonstrate the capacity for balancing the EH&S objectives with socioeconomic development in individual decisions. These positive findings are all the more significant in that they contradict the commentary of at least some observers about pervasive and continuing weaknesses in Poland's EH&S regulatory system (Klarer and Francis 1997; Stodulski 1999; French 1990; Fischhoff 1991). Our findings on regulatory practice are supported by the limited amount of standardized information available on environmental conditions and worker safety in Poland. The fatal accident rate in workplaces has been improving, albeit slowly, and indicators of environmental quality show considerable progress that appears to go beyond the short-term gains associated with plant closure and other aspects of economic dislocation.

This is not to deny that more progress is still needed, or that EH&S regulation could be strengthened by greater capital investment, by closing legal loopholes, and by increased resources for inspection and other regulatory activities. But the *direction* of EH&S regulatory reform in Poland is surely positive in terms of improvements in environmental and occupational protection in the context of economic growth. There are also important differences between the environmental and occupational protection systems. In the environmental protection arena, the past has been a source of innovative ideas and institutional resiliency; in the occupational protection arena, it has created barriers to innovation and change. We suggest that the explanation for the similarities and differences in the evolution of the two systems can be found in their respective social histories and in their particular locations in the broader societal context.

As our research has developed, Poland has become the focus of increasing international scholarly and policy interest. Once of concern for its environmental problems, and to a select set of scholars of East and Central European "economies in transition," Poland is now drawing broad and widespread attention for the level of economic and political progress that has been achieved over the past decade. At the end of the first decade of the transition to democracy and market economy, Poland is widely considered an economic and political success. The decentralization of political power, a series of free elections, absence of ethnic strife have laid the foundation for a stable democracy, and an impressive sustained economic growth of over 6% per year (Ernst 1997; Shleifer 1997) lead some analysts to call Poland a new tiger of central and eastern Europe. By 1997 approximately 70% of employment in industry was in the private sector, over half of which was in small and medium-size firms employing 50 to 1,000 people (Ernst 1997; GUS 1998a), and the spirit of entrepreneurship is thriving. The benefits and costs of this economic development have not fallen evenly on the population, however, and the distribution of wealth has become more polarized, and the social welfare system in such key areas as healthcare, education and employment security has been eroded.

This economic and political success has put important limits on the challenge faced by the EH&S system in Poland. Contrary to the initial concerns of some observers, political and economic restructuring has not been undermined by residual post-communist disregard for the rule of law or for the legitimacy of regulatory authorities. Active public support for environmental issues has waned somewhat in the face of concerns over employment and other social welfare issues, but there has not been a major reversal of course, and environmental and occupational protection are still viewed as legitimate societal goals. There has been considerable positive continuity in institutions and in the use of policy instruments. Accordingly, in the case of Poland, broad challenges to the legitimacy of government regulation or the desirability of environmental protection seem unlikely at this time. What remains unclear is how durable the current EH&S regulatory approach might be in the face of a significant and sustained economic crisis, such as that experienced in Southeast Asia in the late 1990s or by Mexico earlier in the decade.

ASSESSING THE EFFECTIVENESS OF THE EH&S SYSTEM IN POLAND

Our goal in this concluding chapter is to evaluate the effectiveness of the environmental and occupational systems in Poland using the typology proposed in Chapter 1 and to explain the findings in the context of Poland's political, social and institutional culture and traditions. We then apply these findings to the ongoing debate about environmental regulatory reform in the United States and other OECD economies. We are interested both in characterizing Poland's EH&S system in terms of the six dimensions of regulatory practice identified in the typology and assessing whether these aspects of regulatory practice contribute significantly to Poland's ability to improve EH&S performance in the context of other societal goals. We begin with the first element of the typology, namely, the degree to which a regulatory system provides clear and predictable expectations with regards to performance, procedures and consequences of firms, regulators and other participants.

Clarity of Expectations

One of the key characteristics of the EH&S system in Poland is its extensive codification of regulatory procedures. Whether operating an existing facility or planning to establish a new one, industrial managers are presented with a consistent and relatively stable set of codified procedures for obtaining environmental operating permits, which, in turn, specify performance expectations, including monitoring and reporting requirements, performance standards, pollution control technology and fees for pollution and resource use. Licensing and enforcement decisions rely on media- and chemical-specific ambient quality standards. The occupational protection system similarly draws on exposure standards and, additionally, provides detailed specifications of control and monitoring technologies. The environmental permitting process is also unambiguous about the consequences of non-compliance and about the roles of the key actors in the regulatory process: the regional authorities, the firms and the technical experts who prepare permit applications and serve as mediators between firms and the authorities on matters of risk, environmental impact and technological feasibility.

Although several analysts have been critical of the permitting system on the grounds of inflexibility, excessive stringency of ambient standards and ill-calibrated pollution charges (see, for example, Anderson and Fiedor 1997 or Bluffstone and Larson 1997), we have repeatedly noted that firms find the clear performance and procedural expectations helpful in incorporating EH&S concerns into their business decisions, independently of their opinions about the fairness of the imposed performance expectation, norms and procedures.

Poland's approach to defining performance expectations, rich in technical analysis and data hungry, was put in place during the communist era, and by all indications receives widespread support among all actors. This approach reflects the high regard for scientific analysis in Polish society. While some policy leaders in Poland point to potential benefits of eliminating some of the detailed technical specifications in the workplace (see Chapter 3), the recent reforms have increased rather than decreased the prescriptive, highly technical nature of the system.

The predictability of implementation has a mixed record. In the environmental arena, where enforcement rules are quite clear and consistent, the rigidity of the procedures for permitting leads to a sanctioned system of case-specific implementation. The occupational arena is dominated by a large number of very strict standards and a lack of formal procedures for negotiating incremental implementation. Here the response of the inspectors can range from the imposition of strict fines to an open disregard of visible violations of the Labor Code. The regulatory response is much less predictable. Not surprisingly, there appears to be greater satisfaction on the part of both businesses and regulators with implementation practices in the area of environmental protection, as compared to that of occupational protection.

The difference in the predictability of enforcement actions of environmental and occupational policies illustrates a fundamental feature of any regulatory system engaged in the simultaneous pursuit of economic growth and EH&S improvements: the utilization of some mechanism(s) for balancing competing social objectives. Balancing can take various forms and can occur at different stages of the policy-making and implementation cycle, ranging from accounting for costs and feasibility in setting exposure standards, to adopting explicit formulas for case-specific decisions, to according discretionary powers to enforcement agents (accompanied by instruments for accountability and punishment). The choice of approach depends on policy objectives, regulatory and social traditions and the judgment of policy makers, but the need for building in such balancing mechanisms is absolute. The fewer and less flexible such mechanisms are, the more informally the tradeoffs will need to be made, often leading to lack of transparency and vulnerability to pressures. Ultimately, such informality can undermine the benefits of policy instruments that provide for clear performance expectations, such as exposure or performance standards, fees and fines, inspection regiments and others, by creating disincentives to compliance and weakening regulatory authority. The occupational protection system, with its strict risk-based standards and lack of formal procedures for negotiating incremental implementation and enforcement, is a clear case of forced informality at the enforcement stage.

We trace these shortcomings in the occupational protection system to its inability to break free from two aspects of its communist past—namely, a highly hierarchical and centralized system of policy making and implementation and an ideological aversion to explicit acknowledgment of any tradeoffs involving workers' health and safety. We return to this topic later in discussing the system's capacity to learn and profit from change.

Appropriate Policy Instruments

Clear and consistent expectations of firms and regulators must be supported by appropriate policy instruments and regulatory practices. Some of the problems we have identified in the occupational protection system in Poland are at least partly the result of the inappropriate policy instruments. Given the stringency of occupational standards, the enforcement arm needs more explicit discretion and guidance for balancing competing objectives in specific decisions. To some extent, shortcomings in this area are compensated for by other regulatory practices, such as frequent inspections of facilities, extensive monitoring of chemical concentrations and other hazardous agents and institutionalized technical assistance provided to firms by the enforcement authorities. These instruments are effective for two reasons. First, they build on a unique Polish resource: the extensive network of local health departments (SANEPIDs) and regional health departments and labor inspectorates. Second, they draw on distinctive characteristics of Poland's occupational regulatory system, namely, the extensive and detailed knowledge the authorities typically have about the firms under their jurisdiction and a shared preference for negotiated problem solving over confrontation. The other policy instrument introduced in the reform process—employee and union driven committees for occupational safety and hygiene—is, in our view, quite appropriate. The objective in establishing these committees was to break apathy and increase workers' participation in their own safety and health protection. The relatively modest impact of these committees on workers' behavior is more the result of the slow pace of cultural change among the labor force than of any misdirection in regulatory initiatives.

In short, the occupational protection system relies on policy instruments whose appropriateness to the needs of policy makers and managers varies. Notably, however, despite a steep decline in support for the occupational protection system among organized labor, and despite the declining role of the state in overseeing all aspects of the production process, fatal occupational accident statistics in Poland have not changed since 1989. This suggests that there are significant compensating mechanisms at work. These become apparent when we discuss information richness and case-specific implementation of EH&S requirements in Poland.

In contrast to the occupational protection system, policies and policy instruments for environmental protection were widely debated during (and even before) the reform period. Market-based tools such as pollution charges and fees for non-compliance, which predictably failed in the centrally planned economy, make sense now, even if economists do not agree on their effectiveness in pollution prevention (for a discussion about the effects of increasing or decreasing the fees see, for example, Anderson and Fiedor 1997; Bluffstone and Larson 1997; Nowicki 1997). Judging by the richness of the debate over the appropriate use of integrated pollution control, pollution trading schemes, pollution fees, bubble concepts, technology-based regulations, air dispersion models, ambient quality standards and performance standards and negotiated compliance schedules, it appears that the environmental protection system in Poland is well poised to make optimal use of its current mix of market-based and command-and-control policy instruments.

When viewed from the context of OECD and EU policies, the most glaring gap in the menu of policy instruments used in Poland is the absence of any formal process to release disaggregated information to communities, NGOs and other groups. As we discuss below, Poland's EH&S system is in other respects relatively information-rich. Disaggregated information disclosure is now an important priority within the U.S. and OECD economies, involving, for example, detailed information on the use of toxic chemicals within local areas. We return to the issue of whether such a program would strengthen the regulatory system in Poland later in this chapter.

Information Richness

Information richness is a prominent feature of the EH&S system in Poland. With respect to facility-specific decisions it consists of several elements: general technical knowledge and familiarity with formal and informal EH&S policies and policy instruments; knowledge about individual corporate facilities, including their technology, management systems, performance record and environmental impacts, financial status, market situation and community role; knowledge and experience that corporate and regulatory professionals have of each other, including their shared history and past behavior, their understanding of each other's values and perceived trustworthiness and of the external constraints on each other's actions; and data on compliance with standards, environmental discharges, exposures to occupational hazards and serious workplace accidents.

Poland's is a demanding regulatory system requiring a considerable amount of data to be submitted by firms and to be processed by the authorities in order to issue environmental permits, calculate environmental fees and fines and make case-specific implementation decisions. The process of generating and using that information draws in regional environmental authorities, local governments (mostly with new facilities), regional and local occupational authorities, independent technical consultants and EH&S professionals at the firms. The custom of making case-specific implementation decisions based on individual reputations and long-lasting working relationships adds to this information pool. Information richness within the Polish EH&S system is as much about developing knowledgeable relationships among key actors as it is about gathering technical data.

Information richness also means generating information of the kind and quality that can be used in making decisions. In our case studies, we observed that extensive information about the facilities and regulatory intent was used in making facility-specific decisions. But we also note that the system could soon become overloaded by the high data demands. Shortages of inspectors, monitoring equipment and technical expertise to gather and interpret data are evident at all levels, and especially in relation to small firms.

The information rich atmosphere is possible in Poland because of several structural characteristics of the EH&S regulatory system. Of particular importance is the existence of a wide network of institutions with community-level outreach. Other significant ingredients include: the firms' willingness to seek technical assistance from occupational enforcement personnel; the relative openness of firms to supplying environmental performance statistics to governmental inspections; and general avoidance of confrontation in cases of conflicting objectives. Notably, most of the extensive institutional network (and firms' willingness to share performance data with, and seek technical assistance from, that network) is the product of the communist era and is clearly continuing in the new social order. Later in this chapter we return to the question of information richness. In particular, we address its social role in facilitating interactions among key actors and the generalizability of this concept across countries and EH&S systems.

Case-Sensitive Implementation

The capacity for case-specific implementation is another prominent feature of the EH&S system in Poland. Here, we refer to the capacity of regulators and firms to utilize information about each other within the context of policy directives in order to achieve socially optimal decisions. Our case studies uncovered a wide spectrum of case-sensitive decisions on a local and regional administrative level, from imposing harsh permit requirements or fines, forcing technological change or even plant closure, to adopting negotiated enforcement schedules, delaying permit decisions for plants that are clearly unable to meet them or even overlooking violations of some occupational safety standards. Some of these decisions followed formal guidelines, such as scheduled enforcement of environmental performance standards or trading fines for investments into pollution control technology; others were entirely informal.

Our research indicates that case-sensitive flexible implementation is a promising approach to regulating industrial pollution in Poland. It would be even more promising, especially in the occupational protection arena, if a richer menu of policy instruments allowed such case-specific decisions to be made more explicitly. The regulators and industrial managers in our study all expressed an opinion that case-sensitive implementation leads to greater net public good than centrally prescribed decision rules would.

Several authors have argued in favor of case-sensitive environmental decisions in Poland, usually on the grounds of reducing their social costs (see, for example, Bluffstone and Larson 1997; Bell and Bromm 1997). Our findings not only support these views but explain *why* case-sensitive implementation is effective in Poland. First, information richness and fairly well explicated performance expectations equalize the negotiating positions of both sides, despite the visible absence of such advocacy groups as organized labor, environmental organizations and the public. Second, firms and regulatory authorities appear to agree on several key points: that EH&S decisions require balancing competing objectives; that confrontation is not an effective way for making good public decisions; that environmental and occupational protection are fundamental goods that merit protection; that public policies intended to protect health and the environment are legitimate and necessary. Having agreement on these key points reduces the level of conflict and creates incentives for seeking mutually acceptable or tolerable solutions. Third, the well-established working relationships among the parties, the knowledge of each other's past record and trustworthiness and the extensive knowledge of the facilities in question collectively serve to "screen out" decision problems where case-sensitive decision making would be most likely to fail—due, for example, to the untrustworthiness of either party or to gross misjudgment of the technical or economic merits of the case.

That is not to say that confrontations never occur or that firms meekly acquiesce to regulatory authorities. The case of the Radom Leather Tannery and the heavy volume of appeals by firms to the ministry show this is not the case. However, we interpret these data as evidence of the systems' capacity to accommodate confrontation and to provide mechanisms for its resolution. They also suggest that the non-confrontational approaches to regulatory decisions are strategic choices by their participants rather than a structural necessity.

Obviously, case-sensitive decisions made in an informal non-transparent manner, such as disregarding violations of occupational standards or purposeful delaying the issuance of environmental permits, are vulnerable to corruption, co-option or errors in judgment. However, our research indicates that regulators' informal decisions are not made in a normative vacuum. Rather, they appear consistently to be driven by three implicit principles: to push firms to the highest level of pollution control that is technologically and financially feasible; to prevent acute threats to public health and the environment regardless of the cost to the firms or the economic loss to the community; to reward firms demonstrating a commitment to environmental improvements with more flexible treatment. It also appears that firms understand these decision criteria. This shared understanding—another element of information richness—would partly compensate for the imprecise performance expectations and suboptimal set of policy instruments in the occupational protection system.

Capacity to Learn and to Profit from Change

The environmental protection system in Poland responded to the new economic and political order with vigorous debate and experimentation, especially regarding the appropriate use of command-and-control and market-based policy instruments, the proper role for information disclosure about industrial performance, the roles of central, regional and local governments, financing environmental protection and the merits of public access to information. Among the more innovative initiatives were: policies regarding reduction of sulfur dioxide emissions, protecting large unpolluted northeastern land areas from future industrialization and economic development (the Green Lungs of Poland program), incorporating environmental liability and cleanup issues into the privatization procedures, increasing flexibility in facility-specific decisions, empowerment of local authorities to participate in environmental decisions about citing new facilities, public disclosure of the list of the worst polluters and creating the national and regional environmental funds.

Some of these initiatives were clearly a response to international pressures and the desire to join the European Union. This is illustrated by the incorporation of the concept of sustainable development as the guiding philosophy of the amendments to the Environmental Protection and Development Act, the heated debate over the public's right to know and the sulfur dioxide policies. Other initiatives represent selective experimentation with approaches proven effective in OECD countries or approaches considered and abandoned in Poland under the centrally planned economy. Others, such as the national fund for environmental protection, represent uniquely Polish approaches to specific post-1989 circumstances.

Debate and policy experimentation in the environmental protection system represent a continuation of a process which was initiated and nurtured since the mid-1970s by what Sabatier refers to as a policy coalition: a group of actors representing different institutions but sharing certain values, belief systems, goals and problem definition (Sabatier 1993; Jenkins-Smith and Sabatier 1993). According to Sabatier's "advocacy coalition framework," which he uses to explain how policy learning and change occur over time, such coalitions are stable over long time periods. They become mobilized into action by changes in various external parameters, such as socioeconomic conditions, the intensity and nature of the problem, systemic governing coalitions and others.

In Poland the environmental advocacy coalition started forming during the 1970s, consisting mostly of academics and other independent scholars. During the 1980s it increased both in size and diversity, incorporating also professional bureaucrats, members of the ruling party apparatus and environmental advocates. During the peak years of Solidarity-led anti-regime struggle, various external elements joined the coalition for political expediency, only to abandon it shortly after 1989. The core of the group remained quite stable during the period of societal transformation, despite changes in the institutional affiliations of its members. That core was instrumental in implementing the reform program.

Three trends are striking about the post-1989 experience in environmental policy learning and change. First, many within the environmental policy community clearly entered the transition period with a fairly well-formulated agenda. Second, the policy advocacy coalition had the capacity to take advantage of what Kingdon (1995) terms a "window of opportunity" and Balcerowicz refers to as time of "extraordinary politics" that briefly accompanies great societal transformations (Balcerowicz 1995, p. 161). They acted quickly, decisively and consistently at a time when the bureaucracy, the public and the fledging private sector were most receptive to change. Third, in implementing the reform agenda, policy leaders were able to build on the past experiences, accumulated knowledge and perceived strengths of the existing system, while responding to the opportunities created by the emergent market economy and democracy. They also showed sensitivity to timing. For example, despite the failure of their initial efforts to implement pollution trading schemes in the energy sector, they advanced the issue again several years later, achieving at least partial success (Andersson 1998).

For all its vitality and demonstrated capacity to learn and profit from change, the environmental advocacy coalition in Poland depends on a relatively narrow set of policy and intellectual elites. As we noted in Chapter 2, little effort has been made to engage industry, regional and local administrators and local governments in the recent policy debates. It is therefore particularly notable that these neglected yet crucial participants in facility-level decisions have responded so well to the recent reforms. Our study suggests that the industrial managers have incorporated the new procedures into their business strategies with ease. The environmental inspectors interviewed in our case studies have clearly responded to the strengthening of the enforcement arm with renewed confidence and vitality.

The occupational protection system in Poland has shown less capacity to learn and profit from change. This is true for both the central administration and the potential beneficiaries of the system: the workers and their labor representatives. Despite evidence that regional and local inspectors have responded well to their increased power, and are well prepared to implement more flexible enforcement tools, policy leaders have resisted open debate of their two central policy instruments: the strict, health-based occupational standards and the absence of formal mechanisms for flexible case-sensitive enforcement decisions. There are also inconsistencies in the institutional attitude towards delegating responsibility to employers and workers: although several changes in the Labor Code specifically aim at this objective, the policy leadership still justifies the strict occupational standards on the grounds that an extra margin of safety is needed because the standards are likely to be regularly violated.

This bureaucratic behavior is partly justified by the apparent attitudes of workers and organized labor. After decades of conditioning to be the passive recipients of the state's policies regarding occupational protection, workers show little interest in taking the initiative for their own protection. And in the new market economy both workers and organized labor view occupational protection as a threat to their employment security and compensation. Clearly, the established work culture has not yet adapted to the new social circumstances. As with all cultural changes, this one will be slow, most likely only emerging with generational turnover. The new generation of workers, whose formative experience will be in Poland's open and entrepreneurial society, are likely to show more concern for their quality of life, including health protection, than the current workforce. The attitude of the labor movement may follow, especially with increasing economic stability of the private sector.

Another explanation of the resistance of the occupational protection system to change is the lack of tradition of debate and policy experimentation. In this area the social histories of the environmental and occupational systems are markedly different. While the environmental protection system was the product of a gradual accumulation of laws and policies and institutional growth and change—all in the context of historical traditions of nature conservation—the occupational protection system in Poland was constructed according to an ideologically driven blueprint designed by the communist state. The institutions, laws, policies and technological specifications down to the most minute detail were all produced by a central state bureaucracy and handed down to the regions and localities for implementation. Furthermore, since the objectives of the policies were clearly laudable and the means of their implementation were owned by the state, there was neither a need nor opportunity for development of an independent constituency that would debate and push for policy experimentation the way the environment-centered independent intellectual elites have done over the decades.

The lack of a tradition of debate and consultation is best illustrated in the mode of introducing the current reforms. As before, the ideas and initiatives are largely generated within the central bureaucracy, which is reluctant to open the system to wider participation, much less make strategic efforts to increase the number of participants by, for example, funding independent think tanks or creating open fora for debating policy alternatives.

Ownership of the System

In Chapter 1 we argued that the key participants in EH&S regulation need to have a stake in the success of the regulatory system and need to recognize that systemic failure would have consequences inimical to their own interests. Both corporate and administrative decisions about compliance with EH&S regulations should be defensible not only in terms of their consequences for the firm or the community but also in terms of their consequences for the system as a whole. We proposed that when the several institutions comprising the EH&S system have made considerable investments in its development, have considered and rejected possible alternatives and have by their actions demonstrated an active interest in how the system operates and what it produces, the system enjoys multiple "ownership."

Our research uncovers some key elements of such ownership. Both the case studies and the survey indicate that industrial managers, central institutions and regional and local authorities have bought into the underlying philosophy of the EH&S system in Poland: that environmental and occupational protection and socioeconomic development need to be pursued simultaneously and, if necessary, balanced with each other. Furthermore, while we have plenty of evidence of inadequate compliance, differences of opinion and even open confrontation, in the aggregate both the firms and the authorities accept the fundamental legitimacy of the regulatory process in Poland and want it to be preserved.

However, these attitudes are not universally shared among all the key actors. First, organized labor views EH&S protection as a threat to economic growth and employment, preferring to see growth now and EH&S protection later. Second, while the general public shows considerable support for these policies, the key beneficiary of the occupational protection policies—the workers—show little interest or support for that system. Third, the environmental protection system additionally has support from independent intellectual elites, many of whom have been the architects of both the pre-1989 and the post-1989 reforms and thus invested heavily in its success. There no such equivalent group of stakeholders relative to the occupational protection system. As discussed in the previous section, the origins of that difference can be found in the historical development of each regulatory system. As a result, the environmental protection system is more resilient to any future pressure and potential challenge from, for example, political opposition or organized industrial sector than is the occupational protection system. Based on the U.S. experience, where we have observed a declining power of organized labor since the 1970s, the ability of unions in Poland to affect the national occupational safety and health policies may continue to be limited as the country's economy shifts from manufacture- and infrastructure-based to service-based (Kuhn and Wooding 1997, p. 45).

EXPLAINING THE EFFECTIVENESS OF POLAND'S EH&S SYSTEM

In Chapter 1 we proposed that one judges an EH&S regulatory system to be successful if it advances EH&S objectives without imposing unreasonable social and economic costs, and does so in ways that enhance rather than undermine the pursuit of other societal goals, such as improvements in social-economic welfare and protection of the rights of individuals. We then identified six dimensions of regulatory practice that likely enhance the capacity of an EH&S system to meet this goal. In the area of environmental regulation, we have found that the regulatory system in Poland matches up well against these criteria. Building upon existing policies and institutions, and after a decade of reforms, environmental regulation in Poland is now a reasonably effective societal institution, with a capacity to pursue EH&S objectives without undue social cost and without interfering with the pursuit of other societal goals. This has been achieved in Poland despite the visible absence of NGOs and other forms of civic engagement in policy implementation.

The occupational protection component, in comparison, has a mixed record. In general, the occupational protection system diverges substantially from our proposed typology of preferred regulatory practice. Unpredictability in enforcement practice generates uncertain expectations for firms. The absence of specific procedures for balancing health and safety and other concerns renders the occupational protection system vulnerable to abuse. And the strengths in such areas as well codified health and safety standards are constrained, especially by the limited ability to learn and profit from change and by limited ownership across societal actors and administrative hierarchies. Therefore, its long-term capacity for pursuing its key objectives without interfering with other issues on the societal agenda is still open to question.

How do we explain Poland's relative success over the last decade in developing the institutional characteristics that contribute to effective EH&S regulation? Three themes appear to be of importance. First, we observed a high degree of positive continuity in institutions, policies and policy instruments within Poland's EH&S system, before and after reform. Second, we found wide sharing of certain values and attitudes among the key societal actors. Three, we noted a broad support for the rule of law and due process.

Continuity can be a source of weakness (e.g., inertia, inability to change and so forth) and of strength (e.g., ability to draw upon accumulated investments and experience). In the context of profound political and economic change in Poland, it might well be presumed that EH&S reform would be marked by little continuity in regulatory practice. And in some areas this has certainly been the case, as in the qualitative shift in enforcement practices in environmental protection. But in many other areas policies and practices that were ineffective in preventing environmental degradation under the external constraints of communist rule have emerged as important resources within the privatized market economy. These resources range from a system of environmental fees and fines instituted early in the period of communist rule to the highly developed infrastructure of local health inspectorates (SANEPIDs). To be sure, incremental reforms have introduced new policy instruments, de-emphasized some old ones and reorganized some agencies. But the fundamental structure of the system and its personnel has been preserved. This continuity has allowed Poland to draw upon considerable accumulated technical and administrative experience in implementing EH&S reforms.

Beyond specific institutions and policy instruments, there is also a broader level of social continuity within the EH&S systems before and after reform. This is well illustrated by a continuing preference for negotiation over confrontation in the conduct of social transactions. Other elements include close interactions among firms, regulatory authorities and independent technical experts and the value placed upon building dependable and predictable working relationships over time and maintaining reputations and trustworthiness in relation to other parties. Roney (1997) has noted a similar emphasis on maintaining social networks in her case study of working culture in a medium-sized, financially successful metal parts manufacturer in Poland. Andersson (1998) has also singled out continuity in institutional arrangements and policies as one of the main reasons for Poland's success with invigorating its environmental protection system.

A second theme explaining Poland's success is that the three groups of actors most active in EH&S policy making and implementation—industrial managers, regional and local government officials and national policy makers—appear to share with each other two fundamental values and attitudes towards EH&S: that public policies aimed at protection of environmental and occupational health are necessary and legitimate and that EH&S decisions require balancing multiple objectives for the common good. Our finding of pro-environment attitudes is not surprising. As we discussed in Chapter 2, these attitudes have a long and rich history in Polish society. With regard to occupational protection, at first glance this claim does not appear to be well supported by the empirical evidence. In Chapters 4 and 5 we presented evidence for inadequate compliance with safety standards by managers, authorities and workers alike and the willingness of workers to trade health protection for compensation. However, upon closer inspection we find more consistency between the sentiments expressed by managers and bureaucrats and their behavior. Workers and organized labor are, indeed, rather indifferent toward safety and health issues. However, both regulatory authorities and firms practice strategic choices in implementing the occupational protection policies by prioritizing implementation decisions according to the nature of hazards and the level of risk. Tolerance of hazards that may cause life-threatening or chronic diseases is much lower than towards less serious health effects or low probability accidents.

How stable are the attitudes towards EH&S protection in the rapidly changing context of Polish society? In a recent comparative study of British, Norwegian and Polish entrepreneurs, Obłoj and Kolvereid (1996) noted that in deciding to start their own firms Polish businessmen were much more motivated by profit than their counterparts in Norway and Britain. Others were similarly critical of the "cowboy style" capitalism emerging in Poland (see, for example, Perdue 1995; Wiśniewski 1996; Horst 1997; Brada, Hess and Singh 1996). Such a profit-oriented entrepreneurial culture would tend to externalize health, safety and environmental costs. On the other hand, Inglehart's three-decade-long exploration of cultural change in 43 contemporary societies shows that Poland is well on its way towards post-modernist values and their attendant concern for the quality of life, including health and safety (Inglehart 1977a, 1997b).

The third theme emerging from our analysis deals with attitudes towards the rule of law and due process. As we pointed out in Chapters 2, 3 and 6, the EH&S reform program in Poland was premised on a plausible but still untested assumption that the key actors in the EH&S process—the regulatory authorities and industrial managers and executives—would respond to legal and policy changes by changing their behavior. Our study generally provides support for this assumption. In the preceding discussion we partially attributed this finding to the wide sharing of certain core values among the policy makers and to institutional continuity. It appears, in addition, that a fundamental acceptance of the rule of law and due process among the key actors is an important explanatory variable for our findings.

The claim that respect for the rule of law is one of the reasons for the observed response among the bureaucrats and industrial managers to the changes in the legal and policy domains is inconsistent with the views expressed by other authors. Sztompka (1992) and Wedel (1988,

1992; Kolarska-Bobinska 1994; Tarkowska 1993; and others) hypothesized in the early years of the current transformation period that one of the legacies of the four decades of the Soviet rule was a society where collective disregard for official rules was the norm. Kochanowicz (1993) writes that "Polish people are short in respect for the rule of law. In the last 200 years disobedience has been seen as an act of patriotism," while Podgórecki (1994) argued that the communist system affected the very nature and terms of social trust which the rule of law demands. Other commentators echoed that view (see, for example, a review by Veneziano 1997; also Kwaśniewski and Watson 1991; Walicki 1990). In contrast, others spoke to the danger of overstating the durability of the four decades of communist rule (Taras 1995; Smolar 1996; Zamoyski 1987; Kersten 1991; Zuzowski 1993). Taking a longer view of Poland's history regarding the rule of law, Krygier (1995) notes: "We do not know how deep communism went, or how long its consequences will endure. Communism was never engraved on tabula rasa. There was always a society there," while Zamoyski notes: "The Russians never understood the Polish preoccupation with civil liberty and constitutional legality (1987, p. 98). Even Podgórecki (1994) qualifies his earlier concerns about the impact of communism on civil values by noting that Poland is an old society with firmly entrenched social features and that to understand the present one needs to look into Poland's history. Cummings (1994) and Walicki (1990) also stress the long roots of Polish civil law, noting that the communist regime preserved most elements of that legal tradition, while Walicki (1990) and Buchowski (1996) stress the long traditions of civil society in Poland.

In a recently completed history of constitutionalism and participative democracy in Poland, Brzeziński (1998) places their origins in the thirteenth century, which saw the emergence of powerful provincial legislative assemblies (*sejm*) that effectively shared power with the crown. According to this author, political writings in the sixteenth century speak to a maturing political culture and the acceptance of the authority of law by Polish gentry (*szlachta*), which represented close to 15% of the population. By the seventeenth century, Poland was moving towards a parliamentary system, separation of powers, a system of checks and balances and an independent inter-provincial judicial system. The Polish constitution of 1791—the first European written constitution—articulated the central values of Poland's maturing political culture: strong democratic traditions among the gentry, decentralization of power, a system of checks and balances on the government, autonomous judiciary and respect for the rule of law. It also formalized institutions to carry out these values—a decentralized governing system with authority dispersed at provincial and municipal levels, separation of the central power among the king, the legislature and the judiciary, an electoral system for nominating judges and a supreme court to hear cases "against the nation."

The constitution also served as a powerful symbol of nationhood and democratic ideals during the more than a century of foreign domination lasting until the end of World War I. Brzeziński notes that the temporary suspension of constitutionalism during Poland's communist era seems like a brief interlude when viewed against the backdrop of seven centuries of development of democratic ideals and respect for the law and due process. He attributes Poland's smooth transition since 1989 to a Western-style democracy to these traditions developed by Poland's "gentry's democracy."

Brzeziński's explanation is consistent with those of other authors. Zuzowski (1993), Zamoyski (1987) and Cole (1999) all note the tradition of political dissent and citizen's right to justice, which they trace to the legacy of gentry's democracy in Poland, and which set Poland apart from other communist states in Europe. In that historical context, the rapid transition to local governance in Poland, noted by several authors (Jensen and Plum 1993; Regulska 1997; Wollmann 1997; Thurmaier 1994; Hicks and Kamiński 1995; Kirschner 1999), and to a fairly legalistic culture (Letowska and Letowski 1996) are not surprising. The broad acceptance of the legitimacy of EH&S policies which we observed is also consistent with the historical evidence. So it appears that the two opposing views—one seeing a widespread disregard for formal rules and the other stressing the successful transition to an orderly society—are excessive. While Poland is not a highly legalistic culture, we do not find widespread cynicism about EH&S policies and social obligations. Rather, there is considerable support for their legitimacy in the free-market economy emerging in Poland.

In sum, continuity, in both the EH&S institutional domain and in the broader social context, is key to understanding Poland's success in making a transition toward an EH&S system that promises to yield sustained future improvements. The key ingredients of this continuity include: a developed political culture, a durable environmental policy coalition, widely shared views on the preferred mode of conducting societal transactions and acceptance of the bureaucratic, policy and judiciary apparatus as the mediator of competing societal agendas. Its presence reduces conflict and promotes decisions that in the aggregate serve the national goal of a simultaneous advancement of economic growth and EH&S protection. We also discover that among the six criteria describing the system's effectiveness, *information richness* is a characteristic that is interlinked with all others. It is an essential element for defining clear and consistent expectations, for making case-sensitive decisions and for identifying appropriate policy tools. Information richness also enhances the system's capacity to learn and profit from change and, as we demonstrate in the next section, contributes towards building the sense of ownership among various actors.

How generalizable are these finding to other EH&S systems? What are the dimensions of information richness as we move among different EH&S systems and different cultures? What is the role of information richness as a policy tool? How relevant is our finding to the ongoing debate about reforming the EH&S system in the United States, other OECD countries and elsewhere? We address these questions in the next section.

INFORMATION AS A POLICY TOOL

Within the United States and other OECD economies, politicians, policy analysts, environmentalists, the regulated community and regulators are acknowledging that new methods of ensuring environmental progress are needed. As suggested in Chapter 1, one of the motivations behind reform is a desire to enhance the ability of EH&S regulatory systems to pursue environmental and occupational protection within the context of other societal goals. Other drivers of reform include a desire to reduce the actual cost of the regulatory process (especially in regard to litigation and delays), pressure to accelerate performance improvement, and technological innovation as well as broader concerns of pollution prevention and sustainable development.

One measure of the intensity of the ongoing debate about reforming the environmental protection system in the United States is the plethora of reports produced on this subject since the mid-1990s and the remarkable convergence of the emerging ideas (see, for example, Knopman 1996; Environmental Policy Institute 1997; Afsah 1996; The Enterprise for the Environment 1998a; Steinzor 1998a; National Academy of Public Administration 1997). Three broad themes characterize these "next generation" reports. One of themes is about setting clear and measurable environmental goals that communicate directions and priorities for environmental protection to all sectors of society, from the national level all the way down to the local level. The establishment of clear goals is judged to be a critical step in communicating to firms the serious commitment to environmental improvement. There is also considerable consensus about the importance of setting shorter-term performance benchmarks—which can measure progress towards the long-term goals and provide the basis for policy reassessment and course correction.

The second theme has to do with information gathering and utilization as a precondition for setting goals and measuring progress. A common recommendation calls for creating robust information systems on environmental conditions, technological and environmental performance, risks to human health and ecosystems and impacts of social policies. One of the reports, for example, emphasizes six categories of information: indicators of environmental conditions, pollution sources and sinks, human exposure and indicators of human health status, interaction between the economy and the environment, mutual impacts between societal activities and the environment, and metrics of corporate and technological performance (The Enterprise for the Environment 1998). Indicators of industrial performance, including the use of natural resources, environmental discharges, socioeconomic impacts and product performance, are also widely discussed (CERES 1999).

The third prominent theme in the next generation reports is flexibility and place-sensitive environmental management. This theme emphasizes matching decision problems with appropriate policy instruments, sensitivity to local environmental conditions and social needs, balancing competing objectives and involving multiple parties in environmental decisions. A broad range of policy instruments is actively debated, including information disclosure programs, multimedia pollution permits, market-based incentives and performance-based standards. The concept of flexibility also includes technical assistance to small enterprises and tailoring compliance programs to the needs of individual firms and business sectors. Partnerships between industry, NGOs, government and the public are promoted as a promising approach to make flexible and case-sensitive environmental decisions.

We find strong similarities between the priorities identified through the recent national debate in the United States and the typology of an effective EH&S system we advanced in Chapter 1. Specifically, the emphasis on choosing appropriate policy tools, setting clear goals, collecting information about performance and on flexibility and place-sensitive environmental management overlap with four of our criteria: clarity of performance expectations, appropriate policy tools, information richness and case-sensitive implementation. As to the other two criteria we proposed—capacity to learn and profit from change and broad ownership—the intensity of the ongoing debate is the best measure of their presumed importance in maintaining the health of a regulatory system.

The need for information is a central thread linking the three themes of the recent national debate in the United States. It is also a key requirement in our topology of an effective regulatory system. Explicitly, information about industrial performance, technology, the state of the environment and public health and costs is clearly necessary for setting long-term policy goals, assessing policy alternatives and monitoring implementation and impacts. But it is also essential for exercising compliance flexibility, place-sensitive environmental protection and matching policy instruments with decision problems.

One justification for gathering and disclosing environmental information is that it actually leads to pollution reduction. It is hoped that disclosure of data on industrial performance and environmental conditions will create powerful incentives for corporate and governmental agencies to strive for improved performance (Young 1996; Crosby 1999; Finnegan and Sexton 1999). Another way in which access to information should lead to pollution reduction is by uncovering new opportunities for cost-effective innovation in pollution technology and management schemes. Thus, producing data on inefficient use of raw materials should create motivation for improvements, accomplishing reduction of environmental discharges in the process. Finally, it is generally assumed that more information leads to more informed, and thus better, decisions.

In recent years there has been a proliferation of databases on industrial performance and environmental conditions, both in the OECD countries and elsewhere. In the United States, these include the Toxic Release Inventory, Massachusetts Toxic Use Reduction Program, CERES reporting, performance metrics, EPA Environmental Monitoring and Performance Program (EMAP), Global Reporting Initiative (CERES 1999), the recent EPA database on toxic air pollution levels by census track and others. Efforts have been made to expand the coverage and to lower the reporting threshold of Toxic Release Inventory and to create more toxicity data for chemicals in common use (see the review by White 1999). Voluntary partnerships between industry and government, based on information disclosure, have also multiplied, such as the 33/50 Program, Common Sense Initiative, EPA Project XL, Massachusetts Environmental Results Program and others. Empirical evidence in support of the thesis that more information leads to pollution reduction is, unfortunately, limited (see, for example, Tietenberg and Wheeler 1998 for a review of supporting evidence; also, Becker and Geiser 1997). Various authors outright question the effectiveness of information disclosure as a performance driver (Davies and Mazurek 1996).

The second type of argument in favor of gathering and disclosing information on industrial-environmental performance is that it empowers the public to become involved in the decisions that affect it. The proponents of this idea also hope that on a local level such civic environmentalism will lead to more collaborative relationships among the parties (Lipschutz and Mayer 1998; John and Mlay 1999; Murdock and Sexton 1999). This will, in turn, lead to greater legitimacy of decisions and the quality of the underlying science and better tailoring of decisions to the problem at hand, all at a lower cost than would occur through confrontational or coercive approaches. Interest in community-based environmental partnership programs, both nationally and on the state and local level, has been growing during the 1990s. Examples include Community Advisory Panels, Good Neighbor Dialogues, Environmental Results Program, ERP (in Massachusetts) and others (John and Mlay 1999; Murdock and Sexton 1999). Ideas and direct prescriptions for achieving success in such efforts have also multiplied. However, as with the environmental performance data bases, empirical evidence of success of these initiatives is uncertain. First, there is no agreement on what constitutes "successful outcomes" or a good participatory process (Lauber and Knuth 1998). Second, evaluations of public participatory programs are rare (Webler 1995, 1997). One such evaluation covering 29 case studies found limited effectiveness and major challenges ahead (Yosie and Herbst 1998).

Despite the embryonal state of our understanding of the role of information in implementing an effective EH&S regulatory system, there is little doubt that the emphasis on its generation and use will grow in the foreseeable future. Some authors view the current emphasis on information gathering as the logical third phase in the evolution of pollution control policies (Tietenberg and Wheeler 1998). In this perspective the first phase, spanning the 1970s and 1980s, was characterized by almost exclusive reliance on coercive standard-based policy tools. In the second phase, from the mid-1980s to mid-1990s, intense interest developed in market-based policy instruments, such as tradable permits, emission charges, deposit-refunds and others, as complements to the command-and-control approach. The third phase is marked by increased generation, use and disclosure of information as a driver of performance improvement.

This growing emphasis on information calls for a closer examination of its potential roles in policy making and implementation. In general, the current debate views information as a neutral raw material for three types of societal activities: (1) setting goals, formulating policies and monitoring implementation; (2) performing analysis of policy and technological alternatives; (3) identifying new hazards. Its generation, according to this view, occurs through some well-accepted standardized activities involving scientific methods and technology, such as taking measurements, coding, recording and interpreting in accordance with theoretical models. Limited attention has been devoted to the actual process of its generation and use. In particular, the interplay between the process of data gathering and utilization in technical analysis, on the one hand, and policy making and implementation, on the other, is poorly understood.

This mechanical and static view of information severely limits the opportunities for thinking about it as a policy tool. First, neither the policy-relevant information nor technical assessments that utilize it are neutral commodities in the age when they are indispensable resources for assessing policy alternatives and making contentious social choices. A rich empirical and theoretical literature has accumulated over the past three decades on the strategic use of scientific and technical information in environmental controversies by policy makers, scientists and other competing interests (Jasanoff 1990; Nelkin 1995; Kunreuther and Slovic 1996; Porter and van der Linde 1995; Ashford and Rest 1999). It is partly in response to this reality that the Presidential/Congressional Commission on Risk Assessment and Risk Management (1997) and a committee of the National Academy of Sciences have recently called for stakeholder (meaning primarily the public) involvement in all stages of risk assessment, including decisions on what information to collect and how to use it in technical and policy analysis (Stern and Fineberg 1996). Support for stakeholders involvement in data gathering and technical analysis has been growing in other sectors of environmental protection as well (Renn, Webler and Wiedemann 1995a; Stern and Fineberg 1996; Botcheva 1998).

Second, gathering of policy-relevant information hardly resembles turning on some finely calibrated instruments. Rather, it entails sustained interactions among numerous parties with stakes in its content and use. Take, for example collecting data on pollution sources and sinks or on the use and release of toxic materials. It requires identifying pollution sources and the associated human activities, negotiating an agreement on the list of agents, reporting thresholds and technologies of concern, securing disclosure of information by corporations, town officials, communities and other generators of pollution—either voluntarily or through coercion—verification of data and so on. Similarly, in order to gather data on human exposures and health status, the populations at risk must be identified, agreements on diseases of concern must be made, access to health statistics must be negotiated, communities must agree to participate and the methodology for data gathering and analysis must be accepted. Creating information on the interaction between the economy and the environment is even more complex.

Achieving flexibility in policy making and implementation relative to firms, localities or industrial sectors and building partnerships—two other approaches favored by policy reform advocates—present another information challenge. Here, the parties must agree to disclose types of information that in an adversarial atmosphere might be their most valuable weapons and bargaining tools, such as pollution control technologies, optional manufacturing processes and benefits and costs of alternative technologies.

Analysis of the EH&S system in Poland shows that the definition of information richness is context dependent. In the Polish context, information richness also includes the knowledge and experience that corporate and regulatory professionals have of each other: their shared history and past behavior and their understanding of each other's values and attitudes as well as firms' financial status, market situation and community. At the same time, one form of information-richness—namely, a formal program of public information disclosure to NGOs, communities and other groups—has been largely absent in Poland. This is one part of a broader characteristic of regulatory practice in Poland—the limited participation of NGOs and community groups in implementation and enforcement of performance requirements.

When information is viewed as the product of a dynamic interaction between human judgment, advocacy, the scientific method and technology *in a particular societal context*, its potential role as a policy tool expands. It is no longer just an input into analysis of specific policy or technical questions, such as trends in ambient air concentrations of specific pollutants or sources of radioactive waste. Rather, it is a *social process*. The information gathering exercise provides opportunities for new groups and individuals to interact with respect to goal setting and policy making, and possibly reaching common ground. We find various examples of this phenomenon in environmental policy making on national, international and regional scales, such as crafting the national reformulated gasoline policy in the United States, producing the acid rain amendments to the 1990 Clean Air Act (Weber 1998), debating the greenhouse gas policy in the United States (Miller et al. 1998) and the U.S.-Canada Air Quality Accords (Clark and Dickson 1998), development of the tropospheric ozone pollution policy in Europe and the eastern United States (Clark and Dickson 1998), economic analysis of alternative air pollution policies in Poland (Botcheva 1998) and others (see, for example, Bennett 1991; Clark 1997; Fridtjof 1995).

These analyses suggest that data gathering and subsequent technical analysis—the central components of these policy debates—were among the key vehicles for creating a dialogue among parties, along with other more structured efforts to forge partnerships and consensus. Moreover, the broad participation of competing interests in the process of problem definition, data gathering and technical assessment significantly contributed to the success of these efforts, as measured by the wide acceptance of the results and their significant role in policy formulation. On another level, information gathering facilitates interaction among individuals and groups with respect to place-sensitive policy making and flexible implementation. In such cases, it serves as a catalyst for civic environmentalism, for engaging industrial managers, local governments and communities in reconciling competing agendas and for increasing acceptability of decisions. A case of managing pollution from an oil refinery in Virginia, described by Weber (1998a), provides a good illustration. In that case the corporate and government officials agreed to create a comprehensive database on pollution sources, emissions and sinks, which would then be used to jointly develop pollution prevention options and implementation plan. The case highlights the importance of face-to-face contact between regulators and the regulated community, in addition to the familiarity with technical aspects of the case. It also shows how trust developed among the traditional adversaries through joint data gathering and assessment, frequent personal interactions over time and mutual commitment to shared objectives.

The process of information gathering can also contribute to creating a sense of ownership of a regulatory system because it requires a large initial investment from all the participants. One such investment consists of the social cost involved in reaching an agreement about information gathering and then implementing it. The second type of investment is the trust which must develop among the participants before information disclosure is achieved. In the case described by Botcheva (1998) of the conduct of economic analysis of alternative air pollution control policies in Poland, technical experts and academic institutions that participated in the analysis and policy debate developed stakes in the success of these policies and the regulatory system on which the implementation process would depend.

The catalytic effect of information gathering on creating interactions and collaborations and enhancing the sense of ownership in the success of the regulatory system is particularly relevant to the recent policy debate in the United States and elsewhere. There has been growing interest in collaboration as an alternative to the confrontational mode of environmental decision making. Promises of reduced transaction costs to the participants and the society, more optimal technological solutions and greater flexibility in specific decision problems drive these efforts. The "pluralism by the rules" approach advanced by Weber (1998a), civic environmentalism and community-based environmental management (John and Mlay 1999; Murdock and Sexton 1999) are some of the proposed approaches to collaborative decision making. The collaborative approach is, however, a risky proposition. First, the American context provides powerful disincentives to collaboration, such as an adversarial political culture, a fragmented interest group system which expands the agenda and makes reaching consensus more difficult and an open political system which generates multiple opportunities for each party to contest decisions and seek their own agendas through alternative political and legal pathways. Second, it requires an element of trust among parties with competing interests (at least in the short term) and a long history of open confrontation.

The information gathering process can play an important role in strengthening such collaborative efforts by enhancing interactions among the parties, creating opportunities for developing trust and improving the legitimacy of the data thus produced. This is particularly important in case-specific decisions and flexible implementation efforts. There, mutual knowledge among the participants and reputations and attitudes, in addition to such more obvious elements as physical, economic and technical performance measures, play pivotal functions in creating a climate for collaboration.

In Chapter 1 we defined information richness in pollution-related policy making and implementation, recognizing that information gathering and technical assessments are dynamic and interactive social processes. Accordingly, the definition of information richness was inclusive, denoting not only performance indicators but also familiarity with formal and informal EH&S policies and policy instruments and knowledge about individual corporate facilities, including their technology and management system and performance record. Our research in Poland suggests that this definition can be expanded, to explicitly include the human factor. In some social contexts, information richness may also include building lasting relationships among institutions and among individuals and creating an atmosphere of mutual knowledge and predictability of intentions and decisions.

CONCLUSION

Our analysis helps to identify those core elements of a regulatory approach that support improvements in EH&S protection in the context of the pursuit of other societal goals. The first important element is the presence of clear expectations regarding performance requirements, regulatory procedures and consequences of non-compliance. Typically this will include both long-term goals as well as short-term performance benchmarks. Second, the regulatory system must be supported by appropriate policy instruments and approaches. While much of the current debate is about the relative merits of command-and-control policies, market-based instruments and information disclosure, we place emphasis in our work on the fit between policy instruments and prevailing modes of conducting societal transactions such as negotiated problem solving, the role of trust among parties or the provision of technical assistance to firms. Thirdly, we discuss in detail the importance of information richness to the capacity of a regulatory system to accommodate multiple, sometimes competing, goals. We also identify the ability of the regulatory system to learn and profit from change as a key contributor to long-term success.

Few policy makers today would disagree that these four elements of regulation are important to the success and durability of EH&S protection systems. While there would be healthy debate about specific policy instruments, about the amount and type of information needed and other details, the general value of these four elements is now widely acknowledged. The same cannot be said of the remaining two elements, namely, case-specific implementation and the development of broad-based ownership of the regulatory system by all participants. Some regard case-specific implementation, and the associated flexibility in practice and procedures, as too costly or an opportunity for firms to circumvent EH&S performance obligations. Similarly, the concept of broad ownership of the regulatory system by all participants, including firms, is sometimes viewed as placing misguided dependence on the good-faith behavior of firms whose core economic interests lie in avoidance and externalization of environmental and occupational protection costs.

Our case study analysis indicates the importance of developing regulatory systems that include these two dimensions, even if this requires additional investments on the part of regulators and society at large. In short, we have found that case-specific implementation and broad ownership of the regulatory system are critical to the capacity of the regulatory system to balance EH&S objectives and other societal goals, and we advocate the development of regulatory systems in ways that ensure their effective implementation. Thus a commitment to case-specific implementation has important implications for the choice of policy instruments, for such issues as the familiarity of regulators with individual facilities and more generally for the range and type of information generated within the regulatory system. Similarly, the development of broad ownership of the regulatory system is a goal that can only be pursued in context of other dimensions of regulatory practice, such as the consultative processes used to establish performance requirements, the willingness of regulators to respond to case-specific circumstances and the ability of regulators to distinguish between malfeasance and good-faith efforts by firms to achieve improvement within economic and technological constraints. The six dimensions of regulatory practice, in short, intersect within one another in important ways to structure the capacity of an EH&S system to meet societal goals.

One issue of importance is the relevance of Poland's experience and of this typology of regulatory practice to other developing economies. It is fair to say that there are currently two divergent viewpoints on the issue of the generalizability of the regulatory experience of particular countries. The dominant viewpoint supports convergence around a global "best practice" in regulatory approaches, typically involving some version of the regulatory approaches developed within the United States and other OECD economies over the past three decades. Most of the development aid committed to EH&S regulatory improvement in developing countries and in post-Soviet Europe involves transfer of, and training in, regulatory approaches of the OECD. This institutional investment ranges from support for standards-based enforcement regimes, to implementation of market-based regulatory instruments, to fostering NGOs and other elements of public participation (such as community organizing and right-to-know laws). Such transfers of EH&S policy models are increasingly reinforced by the growing importance of regional and international environmental policies, by the actions of large multinational corporations, whose internal environmental management systems apply standardized practices to multiple corporate facilities around the world, and by the growing international significance of standardized perform-

Bibliography

Academy of Social Sciences and Institute for Study of Working Class (1987). *Occupational Hazards in Industry and the Problem of Excess Mortality Among Men*. Warsaw: Polish United Workers Party (in Polish).

Adamson, Seabron, Robin Bates, Robert Laslett and Alberto Pototschnig (1996). *Energy Use, Air Pollution, and Environmental Policy in Krakow: Can Economic Incentives Really Help?* World Bank Technical Paper Number 308, Energy Series. Washington, DC: World Bank.

Adler, Robert W., Jessica C. Landman and Diane M. Cameron (1993). *The Clean Water Act 20 Years Later*. Washington, DC: Island Press.

Afsah, Shakeb, Benoit Laplante and David Wheeler (1996). *Controlling Industrial Pollution: A New Paradigm*. Washington, DC: World Bank.

Allenby, Braden R. and Deanna J. Richard (1994). *The Greening of Industrial Ecosystems*. Washington, DC: National Academy of Engineering, National Academy Press.

Anderson, Glen D. and Boguslaw Fiedor (1997). "Environmental Charges in Poland." In Randall Bluffstone and Bruce A. Larson (eds.), *Controlling Pollution in Transition Economies: Theories and Methods*. Cheltenham: Edward Elgar, pp. 187–208.

Anderson, Terry and Donald R. Leal (1991). *Free Market Environmentalism*. Boulder, CO: Westview Press.

Andersson, Magnus (1998). "Nine Lessons from Poland." *International Environmental Affairs* 10 (1): 3–7.

Andrews, Richard N. L. (1997a). "The Unfinished Business of National Environmental Policy." In Ray Clark and Larry Canter (eds.), *Environmental Policy and NEPA: Past, Present and Future*. Boca Raton, FL: St. Lucie Press, pp. 89–98.

Andrews, Richard N. L. (1997b). "Risk-Based Decisionmaking." In Norman Vig and Michael Kraft (eds.), *Environmental Policy in the 1990s: Reform or Reaction?* Third Edition. Washington, DC: CQ Press, pp. 208–230.

Andrews, Richard N. L. (1999). *Managing the Environment, Managing Ourselves: A History of American Environmental Policy*. New Haven, CT: Yale University Press.

Anonymous, Engineering News-Record (1996). "Intel and EPA Strike a Deal." *Engineering News-Record* 237 (23): 20.

Arora, Seema and Timothy N. Cason (1995). "An Experiment in Voluntary Environmental Regulation: Participation in EPA's 33/50 Program." *Journal of Environmental Economics and Management* 28 (3): 271.

Ashford, Nicholas A. (1991). "Legislative Approaches for Encouraging Clean Technology." *Toxicology and Industrial Health* 7 (5–6): 335–346.

Ashford, Nicholas A. and Charles C. Caldart (1996). *Technology, Law, and the Working Environment*. Covello, CA: Island Press.

Ashford, Nicholas A. and Christopher T. Hill (1980). *The Benefits of Environmental, Health, and Safety Regulation*. Washington, DC: U.S. Government Printing Office.

Ashford, Nicholas A. and Kathleen M. Rest (1999). *Public Participation in Contaminated Communities*. Cambridge, MA: Center for Technology, Policy and Industrial Development, Massachusetts Institute of Technology.

Asian Development Bank (1997). *Emerging Asia: Changes and Challenges*. Manila, Philippines: Asian Development Bank.

Aspen Institute Program of Energy, the Environment, and the Economy (1996). *The Alternative Path: A Cleaner, Cheaper Way to Protect and Enhance the Environment*. The Aspen Institute Series on the Environment in the 21st Century. Washington, DC: Aspen Institute.

Australian Center for Environmental Law (1996). *ISO 14000: Regulation, Trade and Environment*. Canberra, ACT: Australian National University.

Axelrod, Regina (1997). "Environmental Policy and Management in European Union." In Norman Vig and Michael Kraft (eds.), *Environmental Policy in the 1990s: Reform or Reaction?* Third Edition. Washington, DC: CQ Press, pp. 299–320.

Balcerowicz, Leszek (1995). *Socialism, Capitalism, Transformation*. Budapest: Central European University Press.

Barber, Benjamin R. (1984). *Strong Democracy: Participatory Politics for a New Age*. Berkeley: University of California Press.

Bartlett, Robert V. (1997). "The Rationality and Logic of NEPA Revisited." In Ray Clark and Larry Canter (eds.), *Environmental Policy and NEPA: Past, Present and Future*. Boca Raton, FL: St. Lucie Press, pp. 51–60.

Becker, Monica and Ken Geiser (1997). *Evaluating Progress: A Report on the Findings of the Massachusetts Toxics Use Reduction Program Evaluation*. Lowell, MA: Toxics Use Reduction Institute, University of Lowell.

Bell, Ruth Greenspan (1994). "Capital Privatization and the Management of Environmental Liability." In Greta Goldenman (ed.), *Environmental Liability and Privatisation in Central and Eastern Europe*. Washington, DC: World Bank, pp. 108–127.

Bell, Ruth Greenspan and Susan E. Bromm (June 1997). "Lessons Learned in the Transfer of U.S.-Generated Environmental Compliance Tools: Compliance

Schedules for Poland." *Environmental Law Reporter. News and Analysis*: 10296–10303.

Benczek, Krzysztof (1996). *Comparison of Polish and European Union's Occupational Standards*. Internal Document of the Institute of Labour Protection, Warsaw (in Polish).

Bennett, G. (1991). "The History of the Dutch National Environmental Policy Plan." *Environment* 33: 6–9, 31–33.

Bernhard, Michael (1996). "Civil Society after the First Transition: Dilemmas of Post-Communist Democratization in Poland and Beyond." *Communist and Post-Communist Studies* 29 (3): 309–330.

Bertalanffy, Ludgwig von (1968). *General System Theory: Foundations, Development, Applications*. New York: George Braziller.

Bluffstone, Randall and Bruce A. Larson (1997). "Implementing Pollution Permit and Charge Systems in Transition Economies: A Possible Blueprint." In Randall Bluffstone and Bruce A. Larson (eds.), *Controlling Pollution in Transition Economies: Theories and Methods*. Cheltenham: Edward Elgar, pp. 253–270.

Bochniarz, Zbigniew and Richard S. Bolan (1998). "Sustainable Institutional Design in Poland: Putting Environmental Protection on a Self-Financing Basis." In John Clark and Daniel H. Cole (eds.), *Environmental Protection in Transition: Economic, Legal and Socio-Political Perspectives on Poland*. Brookfield, VT: Ashgate.

Bolan, Richard S. and Zbigniew Bochniarz (1994). *Poland's Path to Sustainable Development: 1989–1993*. Minneapolis: Hubert H. Humphrey Institute of Public Affairs, University of Minnesota.

Bonus, H. and H. Niebaum (1997). "Benefits and Costs of Regulating the Environment: Eight Case Studies." *Environment and Planning C: Government and Policy* 15 (3): 329–346.

Botcheva, Liliana (1998). *Doing Is Believing: Participation and Use of Assessments in the Approximation of EU Environmental Legislation in Eastern Europe*. Cambridge, MA: Belfer Center for Science and International Affairs, John F. Kennedy School of Government. http://environment.harvard.edu:80/HERO/wrapper/pageid=gea/geahome.html.

Boyd, James (1998). *Searching for the Profit in Pollution Prevention: Case Studies in the Corporate Evaluation of Environmental Opportunities*. Discussion Paper 98–30. Washington, DC: Resources for the Future.

Brada, Josef C., Alexandra Hess and Inderjit Singh (1996). "Corporate Governance in Eastern Europe: Findings from Case Studies." *Post-Soviet Geography and Economics* 37 (10): 589–614.

Bradley, Theresa (ed.) (1998). *Public Finance Restructuring for Sustainable Development in Emerging Market Economies*. Washington, DC: World Resources Institute.

Brehm, J. and J. T. Hamilton (1996). "Noncompliance in Environmental Reporting: Are Violators Ignorant, or Evasive of the Law?" *American Journal of Political Science* 40: 444–450.

Broszkiewicz, R., B. Krzyśków and H. S. Brown (1998). "Occupational Safety and Health System in Poland during Transition to Democracy and Market Economy." *New Solutions* 8 (2): 221–242.

Brown, Halina Szejnwald, David Angel and Patrick Derr (January/February 1998). "Environmental Reforms in Poland." *Environment* 40 (1): 10–18.

Brown, Halina Szejnwald, Patrick Derr, Otwin Renn and Allen L. White (1993). *Corporate Environmentalism in a Global Economy: Societal Values in International Technology Transfer*. Westport, CT: Quorum Books.

Brown, Halina Szejnwald, Robert Goble and Henryk Kirschner (1995). "Social and Environmental Factors in Lung Cancer Mortality in Post-War Poland." *Environmental Health Perspectives* 103: 64–70.

Brown, Halina Szejnwald, Jeffrey J. Himmelberger and Allen White (1993). "Development-Environment Interactions in the Export of Hazardous Technologies: A Comparative Study of Three Multinational Affiliates in Developing Countries." *Technological Forecasting and Social Change* 43: 125–155.

Brown, Mark B., Weert Canzler and Andreas Knie (1995). "Technological Innovation through Environmental Policy: California's Zero-Emission Vehicle Regulation." *Public Productivity and Management Review* 19 (1): 77.

Brown, Valerie A. (1993). "The Uses of Social and Environmental Health Indicators in Monitoring the Effects of Climate Change." *Climatic Change* 25 (3–4): 389–404.

Bryner, Gary C. (1995). *Blue Skies, Green Politics: The Clean Air Act of 1990 and Its Implementation*. Second Edition. Washington, DC: Congressional Quarterly Press.

Brzeziński, Mark (1998). *The Struggle for Constitutionalism in Poland*. New York: St. Martin's Press.

Buchowski, Michal (1996). "The Shifting Meanings of Civil and Civic Society in Poland." In Chris Hann and Elizabeth Dunn (eds.), *Civil Society: Challenging Western Models*. London and New York: Routledge, pp. 79–98.

Budnikowski, Adam (1992). "Foreign Participation in Environmental Protection in Eastern Europe: The Case of Poland." *Technological Forecasting and Social Change* 41: 147–160.

Bunyagidj, Chaiyod and David Greason (1996). "Promoting Cleaner Production in Thailand: Integrating Cleaner Production into ISO 14001 Environmental Management Systems." *Industry and Environment* 19 (3): 44–47.

Butterworth, Frank M. (1995). "Introduction to Biomonitors and Biomarkers as Indicators of Environmental Change." *Environmental Science Research* 50: 37–53.

Central Institute for Labor Protection (November 1994). "Economic Incentives in the Social Security Payments for Improving the Working Conditions." Prepared for the Ministry of Labor and Social Policy and for the Social Security Administration, Warsaw.

CERES (1998). *Tenth Anniversary Report*. Boston: Coalition for Environmentally Responsible Economies (CERES).

CERES (March 1999). *Global Reporting Initiative: Sustainability Reporting Guidelines. Exposure Draft for Public Comment and Pilot Testing*. Boston: Coalition for Environmentally Responsible Economies (CERES).

Clark, Dick (1997). *The Convergence of US National Security and the Global Environment: Second Conference, March 31–April 5*. Washington, DC: Aspen Institute.

Clark, John and Aaron B. Wildavsky (1990). *The Moral Collapse of Communism: Poland as a Cautionary Tale*. San Francisco: Institute for Contemporary Studies.

Clark, Ray and Larry Canter (eds.) (1997). *Environmental Policy and NEPA: Past, Present and Future*. Boca Raton, FL: St. Lucie Press.

Clark, William C. and Nancy M. Dickson (1998). *The Global Environmental Assessment Project: Overview for 1998*. Cambridge, MA: Belfer Center for Science and International Affairs, John F. Kennedy School of Government. http://environment.harvard.edu:80/HERO/wrapper/pageid=gea/geahome.html

Clift, Roland (1997). "Clean Technology—The Idea and the Practice." *Journal of Chemical Technology and Biotechnology* 68 (4): 347–350.

Coglianese, Cary (April 1999). "The Limits of Consensus." *Environment* 41: 28–33.

Cole, Daniel H. (1995a). "Environmental Protection and Economic Growth: Lessons from Socialist Europe." In Robin P. Malloy and Christopher K. Braun (eds.), *Law and Economics: New and Critical Perspectives*. New York: Peter Lang, pp. 295–329.

Cole, Daniel H. (1995b). "Poland's Progress: Environmental Protection in a Period of Transition." *The Parker School Journal of East European Law* 2 (3): 279–319. New York: Parker School of Foreign and Comparative Law, Columbia University.

Cole, Daniel H. (1997). *Instituting Environmental Protection: From Red to Green in Poland*. New York: St. Martin's Press.

Cole, Daniel H. (1999). "From Renaissance Poland to Poland's Renaissance." *Michigan Law Review* 97 (6): 2062–2102.

Cole, Daniel H. and John Clark (1998). "Poland's Environmental Transformation: An Introduction." In John Clark and Daniel H. Cole (eds.), *Environmental Protection in Transition: Economic, Legal and Socio-Political Perspectives on Poland*. Aldershot: Ashgate, pp. 1–18.

Cooney, Catherine M. (1997). "Emission Inventories under Scrutiny." *Environmental Science and Technology* 31 (9): 406A–407A.

Costanza, Robert, Charles Perrings and Cutler Cleveland (1997). *The Development of Ecological Economics*. Cheltenham: Edward Elgar.

Council on Environmental Quality (CEQ) Annual Reports. *The State of the Environment*. Washington, DC: Executive Office of the President, Council on Environmental Quality.

Covello, Vincent T. (1983). "The Perception of Technological Risks: A Literature Review." *Technological Forecasting and Social Change* 23: 285–297.

Crosby, Ned (1999). "Using Citizens Jury™ Process for Environmental Decision Making." In Ken Sexton (ed.), *Better Environmental Decisions: Strategies for Governments, Businesses, and Communities*. Washington, DC: Island Press, pp. 331–352.

Crowfoot, James E. and Julia M. Wondolleck (1990). *Environmental Disputes: Community Involvement in Conflict Resolution*. Covello, CA: Island Press.

Cruz, Wilfrido, Mohan Munasinghe and Jeremy Warford. (1996). "Greening Development: Environmental Implications of Economic Policies." *Environment* 38 (5): 6–20.

Cummings, Leslie Edwards and N. Joseph Cayer (1993). "Environmental Policy Indicators: A Systems Model." *Environmental Management* 17 (5): 655–667.

Cummings, Susan S. (1994). "Environmental Protection and Privatization: The Allocation of Environmental Responsibility and Liability in Sale Transactions of State-Owned Companies in Poland." *Hastings International and Cooperative Law Review* 17 (3): 551–609.

Dasgupta, Susmita, Benôit Laplante and Nlandu Mamingi (1998). *Capital Market Responses to Environmental Performance in Developing Countries*. Washington, DC: World Bank, Development Research Group.

Davies, Clarence, J. and Jan Mazurek (1998). *Pollution Control in the United States: Evaluating the System*. Washington, DC: Resources for the Future.

Davies, Terry and Jan Mazurek (1996). *Industry Incentives for Environmental Improvement: Evaluation of U.S. Federal Initiatives*. Washington, DC: Global Environmental Management Initiative (GEMI), Resources for the Future, Center for Risk Management.

Davis, Charles (1992). "State Environmental Regulation and Economic Development: Are They Compatible?" *Policy Studies Review* 11 (1): 149–157.

Davis, Charles, E. (1993). *The Politics of Hazardous Waste*. Englewood Cliffs, NJ: Prentice-Hall.

DiSimone, Leon and Frank Popoff (1997). *Eco-efficiency: The Business Link to Sustainable Development*. Cambridge, MA: MIT Press.

Ditz, Daryl W. and Janet Ranganathan (1997). *Measuring Up: Toward a Common Framework for Tracking Corporate Environmental Performance*. Washington, DC: World Resources Institute.

Domański, Henryk (1991). "Structural Constraints on the Formation of the Middle Class." In A. A. Adamski and Edmund Wnuk-Lipinski (eds.), *Challenges to Pluralism in Eastern Europe*. Warsaw: Polish Academy of Sciences.

Douglas, Mary and Aaron B. Wildavsky (1982). *Risk and Culture: An Essay on the Selection of Technical and Environmental Dangers*. Berkeley: University of California Press.

Dowie, Mark (1995). *Losing Ground: American Environmentalism at the Close of the Twentieth Century*. Cambridge, MA: MIT Press.

Durant, Robert F. (July/August 1984). "EPA, TVA and Pollution Control: Implications for a Theory of Regulatory Policy Implementation." *Public Administration Review* 44: 305–315.

Ember, Lois R. (1990). "Pollution Chokes East-Block Nations." *Chemical & Engineering News* 68 (16): 7–16.

The Enterprise for the Environment (1998). *The Environmental Protection System in Transition: Toward a More Desirable Future*. Washington, DC: Center for Strategic and International Studies.

Environmental Law Institute (1998). *Barriers to Environmental Technlogy Innovation and Use*. ELI Project #960800. Washington, DC: Environmental Law Institute.

Environmental Policy Institute (1997). *Environmental Goals and Priorities: Four Building Blocks for Change*. Washington, DC: Environmental Policy Institute.

Environmental Protection Agency (1996). *Current Regulations for Environmental Protection*. Koszalin: Environmental Protection Agency (in Polish).

Ernst, Maurice (1997). "Dimensions of the Polish Economic Transition: The Ingredients of Success." *Post-Soviet Geography and Economics* 38 (1): 1–46.

Farrow, Scott and Michael Toman (1999). "Using Benefit-Cost Analysis to Improve Environmental Regulations." *Environment* 41 (2): 12–15, 33–38.

Feiock, Richard C. and M. Margaret Haley (1992). "The Political Economy of State Environmental Regulation: The Distribution of Regulatory Burdens." *Policy Studies Review* 11 (1): 158–164.

Fiedor, Boguslaw (1998). "Privatizing the Polish Economy: Benefits and Dangers to the Natural Environment." In John Clark and Daniel H. Cole (eds.), *Environmental Protection in Transition: Economic, Legal and Socioeconomic Perspectives on Poland*. Brookfield, VT: Ashgate, pp. 68–80.

Finnegan, John R., Jr. and Ken Sexton (1999). "Community-Based Environmental Decisions: Analyzing Power and Leadership." In Ken Sexton, Alfred A. Marcus, K. William Easter and Timothy A. Burkhardt (eds.), *Better Environmental Decisions: Strategies for Governments, Businesses, and Communities*. Washington, DC: Island Press, pp. 331–352.

Fiorino, Daniel J. (1995). *Making Environmental Policy*. Berkeley: University of California Press.

Fischer, Kurt and Johan Schot (eds.) (1993). *Environmental Strategies for Industry: International Perspectives on Research Needs and Policy Implications*. Covello, CA: Island Press.

Fischhoff, Baruch (1991). "Report from Poland: Science and Politics in the Midst of Environmental Disaster." *Environment* 33 (2): 12–18.

Fischhoff, Baruch, Paul Slovic, Sarah Lichtenstein, S. Read and B. Combs (1978). "How Safe Is Safe Enough? A Psychometric Study of Attitudes Towards Risks and Benefits." *Policy Sciences* 9: 127–152

Frąckiewicz, Lucyna (ed.) (1989). *Demographic-Social Determinants of Health Protection and Social Security*. Siemiatowice: Diament Publishers (in Polish).

French, Hilary F. (1990). *Green Revolutions: Environmental Reconstruction in Eastern Europe and the Soviet Union*. Worldwatch Paper No. 99. Washington, DC: Worldwatch Institute.

Freudenberg, William, R. (1992). "Heuristics, Biases, and the Not-So-General Publics: Expertise and Error in the Assessments of Risks." In Sheldon Krimsky and Dominic Golding (eds.), *Social Theories of Risk*. Westport, CT: Praeger Publishers, pp. 229–249.

Fridtjof Nansen-stiftelsen på Polhøgda (ECON Center for Economic Analysis) (1995). *Integration of Environmental Concerns into Norwegian Bilateral Development Assistance: Policies and Performance: An Evaluation*. Oslo: Royal Ministry of Foreign Affairs.

Funtowicz, Sylvio, O. and Jerome R. Ravetz (1992). "Three Types of Risk Assessment and the Emergence of Post-Normal Science." In Sheldon Krimsky and Dominic Golding (eds.), *Social Theories of Risk*. Westport, CT: Praeger Publishers, pp. 251–273.

Glaser, Rob (1996). "Permits and Promotion of Cleaner Production." *Industry and Environment* 19 (3): 17–27.

Gliński, Piotr (1996). *Polish Greens* (in Polish). Warsaw: The Polish Academy of Sciences, Institute of Philosophy and Sociology (IFiP PAN).

Gliński, Piotr (1998). "Polish Greens and Politics: A Social Movement in a Time of Transformation." In John Clark and Daniel H. Cole (eds.), *Environmental Protection in Transition: Economic, Legal and Socioeconomic Perspectives on Poland*. Brookfield, VT: Ashgate.

Goggin, Malcolm, L., Ann O'M. Bowman, James P. Lester and Laurence J. O'Toole, Jr. (1990). *Implementation Theory and Practice: Toward a Third Generation*. Glenview, IL: Scott, Foresman/Little, Brown Higher Education.

Goldenman, Gretta (1995). "Central European Approaches to Concerns about Environmental Liability During and After Privatization." *International Environmental Reporter* 18: 696–701.

Goldman, Benjamin A. (1994). *Toxic Wastes and Race Revisited: An Update of the 1987 Report on the Racial and Socioeconomic Characteristics of Communities with Hazardous Waste Sites*. Washington, DC: Center for Policy Alternatives (United Church of Christ).

Gomułka, Stanisław (1995). "The IMF-Supported Programs of Poland and Russia, 1990–1994: Principles, Errors and Results." *Journal of Comparative Economics* 20: 316–346.

Gomułka, Stanisław and Jacek Rostowski (1988). "An International Comparison of Material Intensity." *Journal of Comparative Economics* 12 (4): 475–501.

Gottlieb, Robert (ed.) (1995). *Reducing Toxics: A New Approach to Policy and Industrial Decisionmaking*. Washington, DC: Island Press.

Graham, John (1995). "The Environment in Poland: Nature Conservation." *Eastern and Central European Journal on Environmental Law* 1 (1): 33–59.

Grant, Don Sherman II (1997). "Allowing Citizen Participation in Environmental Regulation: An Empirical Analysis of the Effects of Rights-to-Sue and Right-to-Know Provisions on Industry's Toxic Emissions." *Social Science Quarterly* 78 (4): 859–873.

GUS (1996a). *Environmental Protection 1996*. Warsaw: Central Bureau of Statistics (in Polish).

GUS (1996b). *Annual Statistical Report 1996*. Warsaw: Central Bureau of Statistics (in Polish).

GUS (1997). *Annual Labor Statistics*. Warsaw: Central Bureau of Statistics (in Polish).

GUS (1998a). *Annual Statistical Report 1998*. Warsaw: Central Bureau of Statistics (in Polish).

GUS (1998b). *Environmental Protection 1998*. Warsaw: Central Bureau of Statistics (in Polish).

Hajer, Maarten, A. (1995). *The Politics of Environmental Discourse: Ecological Modernization and the Policy Process*. Oxford: Clarendon Press.

Hammond, Allen L. (1995). *Environmental Indicators: A Systematic Approach to Measuring and Reporting on Environmental Policy Performance in the Context of Sustainable Development*. Washington, DC: World Resources Institute.

Hammond, Allen L. (1996). "Environmental Indicators: A Systematic Approach to Measuring and Reporting on Environmental Policy Performance in the Context of Sustainable Development." *International Environmental Affairs* 8 (1): 92–93.

Harrison, Kathryn and George Hoberg (1994). *Risk, Science and Politics: Regulating Toxic Substances in Canada and the United States*. Montreal: McGill–Queen's University Press.

Hawken, Paul (1993). *The Ecology of Commerce: A Declaration of Sustainability*. New York: HarperBusiness.

Heaton, George (1997). "Toward a New Generation of Environmental Technology." *Journal of Industrial Ecology* 1: 23–32.

Heaton, George R., Robert C. Repetto and Rodney Sobin (1991). *Transforming Technology: An Agenda for Environmentally Sustainable Growth in the 21st Century*. Washington, DC: World Resources Institute.

Hicks, Barbara E. (1996). *Environmental Politics in Poland: A Social Movement between Regime and Opposition*. New York: Columbia University Press.

Hicks, James F. and Bartłomiej Kamiński (1995). "Local Government Reform and Transition from Communism: the Case of Poland." *Journal of Developing Societies* 11: 1–20.

Hirszowicz, Maria (1990). "The Polish Intelligentsia in a Crisis-Ridden Society." In Stanislaw Gomułka and Antony Polonsky (eds.), *Polish Paradoxes*. New York: Routledge, pp. 78–91.

Hollick, Malcolm (1993). "Self-organizing Systems and Environmental Management." *Environmental Management* 17 (5): 621–628.

Holling, C. S. (ed.) (1978). *Adaptive Environmental Assessment and Management*. Winchester and New York: John Wiley and Sons.

Horst, Humphrey, A. F. (1997). *Socialism, Capitalism and Transition: With Special Reference to Poland*. Tilburg, Netherlands: Tilburg University Press.

Hubbell, L. Kenneth and Thomas M. Selden (1994). "The Environmental Failures of Central Planning." *Society and Natural Resources* 7 (2): 169–180.

Ikwue, Anthony and Jim Skea (1996). "Business and the Genesis of the European Community Carbon Tax Proposal." In Richard Welford and Richard Starkey (eds.), *Business and the Environment, A Reader*. Washington, DC: Taylor and Francis, pp. 223–240.

ILO 1994. *International Labor Office: Yearbook of Labor Statistics*. Geneva: International Labor Organization.

Indulski, J. A. and R. Rolecki (1995). "Industrialization and Environmental Health in Poland." *Central European Journal of Public Health* 3 (1): 3–12.

Inglehart, Ronald (1977a). *The Silent Revolution: Changing Values and Political Styles among Western Publics*. Princeton, NJ: Princeton University Press.

Inglehart, Ronald (1997b). *Modernization and Postmodernization: Cultural, Economic, and Political Change in 43 Societies*. Princeton, NJ: Princeton University Press.

Institute of Environmental Protection (1990). *Quality of the Environment in Poland*. Warsaw (in Polish).

Jacobson, Elaine Mullaly (1994). *The Theory and Practice of Pollution Credit Trading in Water Quality Management*. Fort Collins: Colorado State University.

Jaffe, Adam B., Steven R. Peterson, Paul R. Portney and Robert N. Stavins (1995). "Environmental Regulation and the Competitiveness of U.S. Manufacturing: What Does the Evidence Tell Us?" *Journal of Economic Literature* 33 (1): 132–163.

Jahn, Detlef (1998). "Environmental Performance and Policy Regimes: Explaining Variations in 18 OECD-Countries." *Policy Sciences* 31 (2): 107–131.

Jasanoff, Sheila (1987). "Cultural Aspects of Risk Assessment in Britain and the United States." In Brandon Johnson and Vincent Covello (eds.), *The Social and Cultural Construction of Risk: Essays on Risk Selection and Perception*. Dordrecht: D. Reidel Publishing.

Jasanoff, Sheila (1990). *The Fifth Branch: Science Advisers as Policymakers*. Cambridge, MA: Harvard University Press.

Jasiński, Piotr (1996). "Pro-ecological Privatization? Ownership Changes and Natural Environment in Poland, 1989–1994." *Communist Economies & Economic Transformation* 8 (3): 335–362.

Jendrośka, Jerzy (1993). "Integrated Pollution Prevention through Licensing Procedures in Poland." In Betty Gebers and Morga Robesin (eds.), *Licensing Procedures for Industrial Plants and the Influence of EC-Directives*. Frankfurt and New York: Peter Lang.

Jendrośka, Jerzy (1996a). "Drafting New Environmental Law in Poland: Radical Change or Merely Reform?" In Gerd Winter (ed.), *European Environmental Law: A Comparative Perspective*. Brookfield, VT: Dartmouth Publishing Company, pp. 367–390.

Jendrośka, Jerzy (1996b). *Environmental Regulatory Framework in Poland: History and Recent Development*. Proceedings of a September 1995 Workshop on Environmental Health and Safety in Private Enterprises in Poland. Warsaw: Central Institute for Labour Protection, pp. 49–87.

Jendrośka, Jerzy (1998). "Environmental Law in Poland, 1989–1996: An Assessment of Past Reforms and Future Prospects." In John Clark and Daniel H. Cole (eds.), *Environmental Protection in Transition: Economic, Legal and Socioeconomic Perspectives on Poland*. Brookfield, VT: Ashgate.

Jendrośka, Jerzy and Wojciech Radecki (1991). "The Environmental Protection Act of 1980: An Overview and Critical Assessment." In Zbigniew Bochniarz and Richard S. Bolan (eds.), *Designing Institutions for Sustainable Development: A New Challenge for Poland*. Minneapolis and Bialystok: Hubert H. Humphrey Institute, Bialystok Technical University.

Jendrośka, Jerzy and Jerzy Sommer (1994). "Environmental Impact Assessment in Polish Law: The Concept, Development, and Perspectives." *Environmental Impact Assessment Review* 14 (2–3): 169–194.

Jenkins-Smith, Hank C. and Paul A. Sabatier (1993). "The Dynamics of Policy-Oriented Learning." In Paul A. Sabatier and Hank C. Jenkins-Smith (eds.), *Policy Change and Learning: An Advocacy Coalition Approach*. Boulder, CO: Westview Press, pp. 41–60.

Jensen, H. T. and V. Plum (1993). "From Centralised State to Local Government: The Case of Poland in the Light of Western European Experience." *Environment and Planning C: Government and Policy* 11: 565–581.

John, DeWitt (1994). *Civic Environmentalism: Alternatives to Regulation in States and Communities*. Washington, DC: Congressional Quarterly Press.

John, DeWitt and Marian Mlay (1999). "Community-Based Environmental Protection: Encouraging Civic Environmentalism." In Ken Sexton, Alfred A. Marcus, K. William Easter and Timothy D. Burkhardt (eds.), *Better Environmental Decisions: Strategies for Governments, Businesses, and Communities*. Washington, DC: Island Press, pp. 331–352.

Johnson, Brandon B. and Vincent T. Covello (eds). (1987). *The Social and Cultural Construction of Risk: Essays on Risk Selection and Perception*. Dordrecht: D. Reidel Publishing.

Judge, William Q. and Thomas J. Douglas (1998). "Performance Implications of Incorporating Natural Environmental Issues into the Strategic Planning Process: An Empirical Assessment." *The Journal of Management Studies* 35 (2): 241–260.

Kabala, Stanley J. (1985). "Poland: Facing the Hidden Costs of Development." *Environment* 27 (9): 6–42.

Kamieniecki, Krzysztof and Alexandra Kuspak (eds.) (1998). *European Union and Environmental Protection: Selected Facts and Analyses.* Warsaw: Institute for Sustainable Development, ISBN 83–85–787–23–2 (in Polish).

Karaczuń, Zbigniew M. (1995). "Policy of Air Protection in Poland." *Water, Air, and Soil Pollution* 85 (4): 2637–2642.

Karaczuń, Zbigniew M. (1997). *Policy of Air Protection in Poland. Research Report #4.* Warsaw: Institute for Sustainable Development.

Kasperson, Roger E., Dominic Golding and Jeanne X. Kasperson (1999). "Risk, Trust and Democratic Theory." In George Cvetkowicz and Ragner E. Lofstedt (eds.), *Social Trust and the Management of Risk.* London: Earthscan Publications.

Kasperson, Roger E., Jeanne X. Kasperson, Christoph Hohenemser and Robert W. Kates (1988). *Corporate Management of Health and Safety Hazards: A Comparison of Current Practice.* Boulder, CO: Westview Press.

Kasperson, Roger E., Ortwin Renn, Paul Slovic, Halina S. Brown, Jody E. Emel, Robert Goble, Jeanne X. Kasperson and Samuel Ratick (1988). "The Social Amplification of Risk: A Conceptual Framework." *Risk Analysis* 8 (2): 177–187.

Kempa, Edward S. (1997). "Hazardous Wastes and Economic Risk Reduction: Case Study, Poland." *International Journal of Environment and Pollution* 7 (2): 221–248.

Kersten, Krystyna (1991). *The Establishment of Communist Rule in Poland, 1943–1948.* Berkeley and Los Angeles: University of California Press.

Kingdon, John, W. (1995). *Agendas, Alternatives, and Public Policies.* Second Edition. New York: HarperCollins College Publishers.

Kirschner, Emil Joseph (1999). *Decentralization and Transition in the Visegrad: Poland, Hungary, the Czech Republic and Slovakia.* New York: St. Martin's Press.

Kjellstrom, Tord and Carlos Corvalan (1995). "Framework for the Development of Environmental Health Indicators." *World Health Statistics Quarterly* 48 (2): 144–154.

Klarer, Jürg and Patrick Francis (1997). "Regional Overview." In Jürg Klarer and Bedrich Moldan (eds.), *The Environmental Challenge for Central European Economies in Transition.* New York: John Wiley and Sons, pp. 1–66.

Kleindorfer, Paul R. and Eric W. Orts (1998). "Informational Regulation of Environmental Risks." *Risk Analysis* 18 (2): 155–164.

Knopman, Debra S. (1996). "Second Generation—A New Strategy for Environmental Protection." Washington, DC: Progressive Foundation, Center for Innovation and the Environment.

Kochanowicz, Jacek (1993). "The Disappearing State: Poland's Three Years of Transition." *Social Research* 60 (4): 821–834.

Koehler, Dinah and Maximilian Chang (1999). "Search and Disclosure: Corporate Environmental Reports." *Environment* 41 (2): 3.

Kolarska-Bobińska, Lena (1994). *Aspirations, Values and Interests: Poland 1989–94*. Warsaw: IFiS Publishers.

Kopp, Raymond J., Paul R. Portney and Diane E. DeWitt (Fall 1990). "Comparing Environmental Regulation in the OECD Countries." *Resources* 101: 10–13.

Kosobud, Richard F. (ed.) (1997). *Market-Based Approaches to Environmental Policy*. New York: Van Nostrand Reinhold.

Kowalski, Jerzy (1996). *Evaluation of Occupational Health and Safety in Poland in 1994*. Report by the Ministry of Labor and Social Policy in Poland. In Proceedings of a Workshop on Environment, Health and Safety in Private Enterprises in Poland. Central Institute for Labor Protection, Warsaw, pp. 1–30.

Kraft, Michael E. (1997). "Environmental Policy in Congress: Revolution, Reform, or Gridlock?" In Norman Vig and Michael Kraft (eds.), *Environmental Policy in the 1990s: Reform or Reaction?* Third Edition. Washington, DC: CQ Press, pp. 119–167.

Krimsky, Sheldon and Dominic Golding (eds.) (1992). *Social Theories of Risk*. Westport, CT: Praeger Publishers.

Kristiansen, Mark (1996). "Incorporating Environmental Law in the Context of Privatization Transactions in Hungary, Poland and Russia." *Administrative Law Review* 48 (4): 627–644.

Krygier, Martin (1995). "The Constitution of the Heart." *Law and Social Enquiry* 20: 1033–1066.

Kuhn, Sara and John Wooding (1997). "The Changing Structure of Work in the United States: Part 2—The Implications for Health and Welfare." In Charles Levenstein and John Wooding (eds.), *Work, Health, and the Environment: Old Problems, New Solutions*. New York: Guilford Press, pp. 31–58.

Kunreurther, Howard and Paul Slovic (May 1996). "Challenges in Risk Assessment and Risk Management." *The Annals of the Academy of Political and Social Science* 545: 116–125.

Kwaśniewski, Jerzy and Margaret Watson (eds.) (1991). *Social Control and the Law in Poland*. New York: Berg Publishers.

Landsberg, J. H., B. A. Blakesley, P. O. Reese, G. McRae and P. R. Forstchen (1998). "Parasites of Fish as Indicators of Environmental Stress." *Environmental Monitoring and Assessment* 51 (1–2): 211.

Lauber, Bruce T. and Barbara A. Knuth (1998). "Refining Our Vision of Citizen Participation: Lessons from a Moose Reintroduction Proposal." *Society and Natural Resources* 11 (4): 411–424.

Lee, Kai, N. (1993). *Compass and Gyroscope: Integrating Science and Politics for the Environment*. Washington, DC: Island Press.

Lester, James, P. (ed.) (1995). *Environmental Politics and Policy: Theories and Evidence*. Second Edition. Durham, NC: Duke University Press.

Letowska, Ewa and Janusz Letowski (1996). *Poland: Towards the Rule of Law*. Warsaw: Scientific Publishers SCHOLAR, Institute of Legal Studies, Polish Academy of Sciences.

Linpinski, Andrzej and J. Otto (1996). "New Polish Mining and Petroleum Legislation." *Journal of Energy and Natural Resources Law* 14 (3): 325–345.

Lipschutz, Ronnie D. and Judith Mayer (1996). *Global Civil Society and Global*

Environmental Governance: The Politics of Nature from Planet to Planet. Albany: State University of New York Press.

Lotspeich, R. (1998) "Comparative Environmental Policy: Market Type Instruments in Industrialized Capitalist Economies." *Policy Studies Journal* 26: 85–108.

Luken, Ralph Andrew (1990). *Efficiency in Environmental Regulation: A Benefit-Cost Analysis of Alternative Approaches*. Boston: Kluwer Academic Publishers.

Machlis, Gary E. and Eugene A. Rosa (1990). "Desired Risk: Broadening the Social Amplification of Risk Framework." *Risk Analysis* 10 (1): 161–168.

Maclean, Heather and Lester Lave (1998). "A Life Cycle Model of an Automobile." *Environmental Science and Technology* 32 (13): 322A–330A.

Malone, Charles R. (1990). "Environmental Performance Assessment: A Case Study of an Emerging Methodology." *Journal of Environmental Systems* 19 (2): 171–184.

Marcus, Philip A. and John T. Willig (eds.) (1997). *Moving Ahead with ISO 14000: Improving Environmental Management and Advancing Sustainable Development*. New York: John Wiley and Sons.

Mattoo, Aaditya and Harsha V. Singh (1994). "Eco-labelling: Policy Considerations." *Kyklos* 47 (1): 53–65.

Mazmanian, Daniel A. and Jeanne Nienaber (1979). *Can Organizations Change? Environmental Protection, Citizen Participation and the Corps of Engineers*. Washington, DC: Brookings Institution.

Mazurski, Krzysztof R. (1997) "Poland's National Parks: Problems and Prospects." *Forum for Applied Research and Public Policy* 12 (2): 111–114.

McAdams, James A. (ed.) (1997). *Transitional Justice and the Rule of Law in New Democracies*. Notre Dame and London: University of Notre Dame Press, pp. 185–237.

McGarity, Thomas and Sidney A. Shapiro (1993). *Workers at Risk: The Failed Promise of the Occupational Safety and Health Administration*. Westport, CT: Praeger Publishers.

McGrew, Anthony (1993). "The Political Dynamics of the 'New' Environmentalism." In Denis Smith (ed.), *Business and the Environment: Implications of the New Environmentalism*. New York: St. Martin's Press, pp. 12–26.

Michalos, Alex C. (1997). "Combining Social, Economic and Environmental Indicators to Measure Sustainable Human Well-Being." *Social Indicators Research* 40 (1–2): 221–258.

Mikosz, Jerzy (1996). "Water Management Reform in Poland: A Step Toward Ecodevelopment." *Journal of Environment and Development* 5 (2): 233–253.

Milbrath, Lester W. (1984). *Environmentalists, Vanguard for a New Society*. Albany: State University of New York Press.

Milbrath, Lester W. (1989). *Envisioning a Sustainable Society: Learning Our Way Out*. Albany: State University of New York Press.

Millard, Frances (1998). "Environmental Policy in Poland." *Environmental Politics* 7 (1): 145–161.

Miller, Alan S. (1995). "Environmental Regulation, Technological Innovation, and Technology-Forcing." *Natural Resources and Environment* 10 (2): 5–21.

Miller, Clark, Sheila Jasanoff, Marybeth Long, William Clark, Nancy Dickson, Alastair Iles and Tom Parris (1998). *Working Group 2—Assessment as a Com-*

munications Process. Cambridge, MA: The Global Environmental Assessment Project, Belfer Center for Science and International Affairs, John F. Kennedy School of Government. http://environment.harvard.edu:80/HERO/wrapper/pageid=gea/geahome.htm

Ministry of Environmental Protection, Natural Resources and Forestry (1991). *National Environmental Policy of Poland*. Warsaw (in Polish).

Ministry of Labor (May 1996). *Assessment of the State of Occupational Safety and Hygiene in Poland in 1996*. Warsaw: Ministry of Labor and Social Policy (in Polish).

Misztal, Barbara A. (1996). *Trust in Modern Societies: The Search for the Bases of Social Order*. Cambridge, MA: Blackwell Publishers.

Misztal, Bronislaw (1990). "Alternative Social Movements in Contemporary Poland." *Research in Social Movements, Conflict and Change* 12: 67–88.

Mitchell, Robert Cameron (1979). "National Environmental Lobbies and the Apparent Illogic of Collective Action." In Clifford S. Russell (ed.), *Collective Decision Making: Applications from Public Choice Theory*. Baltimore: John Hopkins University Press.

Morgenstern, Richard D. (1997). *Economic Analysis at EPA: Assessing Regulatory Impacts*. Washington, DC: Resources for the Future.

Morgernstern, Richard D. and Mark K. Landy (1997). "Economic Analysis: Benefits, Costs, Implications." In Richard D. Morgernstern (ed.), *Economic Analysis at EPA: Assessing Regulatory Impacts*. Washington, DC: Resources for the Future, pp. 455–478.

Morrison, Charles (1991). *Managing Environmental Affairs: Corporate Practices in the U.S., Canada and Europe*. New York: The Conference Board.

Murdock, Barbara S. and Ken Sexton (1999). "Community-Based Environmental Partnerships." In Ken Sexton, Alfred A. Marcus, K. William Easter and Timothy D. Burkhardt (eds.), *Better Environmental Decisions: Strategies for Governments, Businesses, and Communities*. Washington, DC: Island Press, pp. 331–352.

Nagengast, Carole (1991). *Reluctant Socialists, Rural Entrepreneurs: Class, Culture and the Polish State*. Boulder, CO: Westview Press.

National Academy of Engineering (1999). *Industrial Environmental Performance Metrics: Challenges and Opportunities*. Washington, DC: National Academy Press.

National Academy of Public Administration (1995). *Setting Priorities, Getting Results: A New Direction for the Environmental Protection Agency*. First Edition. Washington, DC: Academy of Public Administration.

National Academy of Public Administration (1996). *Setting Priorities, Getting Results: A New Direction for EPA*. Washington, DC: NAPA.

National Academy of Public Administration (1997). *Resolving the Paradox: EPA and the States Focus on Results*. Washington, DC: NAPA.

Nelkin, Dorothy (1995). "Science Controversies: The Dynamics of Public Disputes in the United States." In Sheila Jasanoff (ed.), *Handbook of Science and Technology Studies*. Thousand Oaks, CA: Sage Publications, pp. 444–456.

Nieuwlaar, E., E. Alsema and B. Van Engelenburg (1996). "Using Life-Cycle Assessments for the Environmental Evaluation of Greenhouse Gas Mitigation Options." *Energy Conversion and Management* 37 (6/8): 831.

Noble, Charles (1997). "OSHA at 20: Regulatory Strategy and Institutional Structure in the Work Environment." In Charles Levenstein and John Wooding (eds.), *Work, Health, and the Environment: Old Problems, New Solutions*. New York: Guilford Press, pp. 125–141.

Novak, Anne M. (1996). "Central and Eastern Europe: U.S. Exports of Environmental Technologies Are Fueled by New Market Economies." *Business America* 117 (4): 50–53.

Nowak, Zygfryd (1996). "Cleaner Production in Poland: From CP to Environmental Management Systems." *Industry and Environment* 19 (3): 12–19.

Nowicki, Maciej (1997). "Poland." In Jurg Klarer and Bedrich Moldan (eds.), *The Environmental Challenge for Central European Economies in Transition*. New York: John Wiley and Sons, pp. 193–228.

Obłoj, K. and M. Kostera (1993). "Polish Privatization Program: Action, Symbolism and Cultural Barriers." *Industrial and Environmental Quarterly* 3 (3): 7–21.

Obłoj, Krzysztof and Lars Kolvereid (1996). "Enterprenuers in Different Environments and Cultures in Britain, Norway and Poland: Towards a Comparative Framework." In Pat Joynt and Malcolm Warner (eds.), *Managing across Cultures: Issues and Perspectives*. London: International Thomson Business Press, pp. 338–359.

OECD (1997). *OECD Environmental Performance Reviews: A Practical Introduction*. Paris: OECD.

OECD (1998a). *Towards Sustainable Development: Environmental Indicators*. Paris: Organization for Economic Co-Operation and Development.

OECD (1998b). *Environmental Performance in OECD Countries: Progress in the 1990s*. Washington, DC: OECD Publications and Information Center.

Offe, Claus (1985). "New Social Movements: Challenging the Boundaries of Institutional Politics." *Social Research* 52 (5): 817–868.

Olsen, Marvin E., Dora G. Lodwick and Riley E. Dunlap (1992). *Viewing the World Ecologically*. Boulder, CO: Westview Press.

O'Riordan, Timothy (1985). "Approaches to Regulation." In Harry Ottway and Malcolm Peltu (eds.), *Regulating Industrial Risks: Science, Hazards and Public Protection*. London: Buttersworths, pp. 20–39.

Ottway, Harry (1985). "Regulation and Risk Analysis." In Harry Ottway and Malcolm Peltu (eds.), *Regulating Industrial Risks: Science, Hazards and Public Protection*. London: Buttersworths, pp. 1–19.

Pargal, Sheoli, Hemamala Hettige, Manjula Singh and David Wheeler (1997). "Formal and Informal Regulation of Industrial Pollution: Comparative Evidence from Indonesia and the United States." *World Bank Economic Review* 11 (3): 433–450.

Pawłowski, Łucjan and Marzenna R. Dudzińska (1994). "Environmental Problems of Poland during Economic and Political Transformation." *Ecological Engineering* 3: 207–215.

Pearce, David W. and R. Kerry Turner (1990). *Economics of Natural Resources and the Environment*. Baltimore: John Hopkins University Press.

Pearce, David W., Edward Barbier and Anil Markandya (1991). *Sustainable Development: Economics and Environment in the Third World*. London: Earthscan Publications.

Perdue, William Dan (1995). *Paradox of Change: The Rise and Fall of Solidarity in the New Poland*. Westport, CT: Praeger Publishers.

Perlez, Jane (1997). "Polish Home That Bison Still Roam." *New York Times*, August 25, p. 1.

Petulla, Joseph M. (1987). *Environmental Protection in the United States: Industry, Agencies, Environmentalists*. San Francisco: San Francisco Study Center.

Podgórecki, Adam (1994). *Polish Society*. Westport, CT: Praeger Publishers.

Pond, Ernest (1998). "Miracle on the Vistula." *The Washington Quarterly* 21: 209–211.

Porter, Michael E. (1990). *The Competitive Advantage of Nations*. New York: The Free Press.

Porter, Michael E. and Claas van der Linde (1995). "Green and Competitive: Ending the Stalemate." *Harvard Business Review* 73 (5): 120–134.

Portney, Paul R. (ed.) (1990). *Public Policies for Environmental Protection*. Washington, DC: Resources for the Future.

Potter, S. and M. Hinnells (1994). "Analysis of the Development of Eco-labelling and Energy Labelling in the European Union." *Technology Analysis and Strategic Management* 6 (3): 73–89.

Poznański, Kazimierz (1994). *Poland's Protracted Transition: Institutional Change and Economic Growth 1970–94*. Cambridge: Cambridge University Press.

Presidential/Congressional Commission on Risk Assessment and Risk Management (1997). *Final Report* (2 vols.). Washington, DC: The Commission.

Pressman, Jeffrey and Aaron Wildavsky. (1984). *Implementation: How Great Expectations in Washington Are Dashed in Oakland*. Third edition. Berkeley: University of California Press.

Prins, R., V.C.A. van Polanen Petel, C. Th. Zandvliet and M. K. Koster (1997). *National Occupational Safety and Health Policies, Their Implementation and Impact: An Eight Country Exploration*. The Hague: VUGA Vitgererij B. V.

Prüfer, Christopher B. (1997). *Waste Management in Poland and Germany: Economic and Legal Aspects of Waste Management and Their Impact on Cooperation at National and Business Level*. Frankfurt am Main: Peter Lang.

Public Law 1965. March 30, 1965 legislation, DZ.U. nr 13, poz. 91 (in Polish).

Public Law 1974. June 26, 1974 legislation on labor code, DZ.U.nr. 24, pox. 141, z pozniejszymi zmianami (in Polish).

Public Law 1981. March 6, 1981 legislation on State Labor Inspectorate, tj. DZ.U.z1985r.nr 54, poz. 276 z pozniejszymi zmianami (in Polish).

Public Law 1983. June 24, 1983 legislation on social labor inspection, DZ.U. nr 35, poz 163 z pozniejszymi zmianami (in Polish).

Public Law 1985. March 14, 1985 legislation on state sanitary inspectorate, DZ.U. nr. 12, poz. 49 z pozniejszymi zmianami (in Polish).

Putnam, Robert D., Robert Leonardi and Raffaella Y. Manetti (March 1983). "Explaining Institutional Success: The Case of Italian Regional Government." *The American Political Science Review* 77: 55–73.

Rabe, Barry G. (1997). "Power to the States: The Promise and Pitfalls of Decentralization." In Norman Vig and Michael Kraft (eds.), *Environmental Policy in the 1990s: Reform or Reaction?* Third Edition. Washington, DC: CQ Press, pp. 31–52.

Rappaport, Ann and Margaret Flaherty (1991). "Multinational Corporations and the Environment: Context and Challenges." *International Environmental Reporter* 14: 261–267.

Rayner, Steve (1992). "Cultural Theory and Risk Analysis." In Sheldon Krimsky and Dominic Golding (eds.), *Social Theories of Risk*. Westport, CT: Praeger Publishers, pp. 83–116.

Regulska, J. (1997). "Decentralization or (Re)centralization: Struggle for Political Power in Poland." *Environment and Planning C: Government and Policy* 15: 187–207.

Renn, Ortwin (1992). "The Social Arena Concept of Risk Debates." In Sheldon Krimsky and Dominic Golding (eds.), *Social Theories of Risk*. Westport, CT: Praeger Publishers, pp. 178–196.

Renn, Ortwin, Thomas Webler and Peter M. Wiedemann (eds.) (1995a). "A Need for Discourse on Citizen Participation: Objectives and Structure of the Book." In Ortwin Renn, Thomas Webler and Peter M. Wiedemann (eds.), *Fairness and Competence in Citizen Participation: Evaluating Models for Environmental Discourse*. Boston: Kluwer Academic, pp. 1–15.

Renn, Ortwin, Thomas Webler and Peter M. Wiedemann (eds.) (1995b). *Fairness and Competence in Citizen Participation: Evaluating Models for Environmental Discourse*. Boston: Kluwer Academic.

Repetto, Robert C. (1995). *Jobs, Competitiveness, and Environmental Regulation: What Are the Real Issues?* Washington, DC: World Resources Institute.

Rider, Christine and Edward K. Zajicek (1995). "Mass Privatization in Poland: Processes, Problems and Prospects." *International Journal of Politics, Culture and Society* 9 (1): 133–148.

Ringquist, Evan. J. (1993). *Environmental Protection at the State Level: Politics, and Progress in Controlling Pollution*. Armonk, NY: M. E. Sharpe.

Roberts, K. H. and George Gargano (1990). "Managing a High Reliability Organization: A Case of Interdependence." In M. A. von Glinow and S. A. Mohrman (eds.), *Managing Complexity in High Technology Organizations*. New York: Oxford University Press, pp. 146–159.

Rondinelli, David and Jan Yurkiewicz (1996)." Privatization and Economic Restructuring in Poland: An Assessment of Transition Policies." *American Journal of Economics and Sociology* 55: 145–154.

Roney, Jennifer (1997). "Cultural Implications of Implementing TQM in Poland." *Journal of World Business* 32 (2): 152–167.

Roome, Nigel (ed.) (1998). *Sustainability Strategies of Industry: The Future of Corporate Practice*. Washington, DC: Island Press.

Rosenbaum, Walter (1997). "The EPA at Risk: Conflicts over Institutional Reform." In Norman Vig and Michael Kraft (eds.), *Environmental Policy in the 1990s: Reform or Reaction?* Third Edition. Washington, DC: CQ Press, pp. 143–167.

Rzepecki, Jan (July 1991). *Differentiation of Premiums Based on the Assessment of Work Conditions as an Economic Stimulus for their Improvement*. Internal Report of the Institute for Labor Protection, Warsaw (in Polish).

Sabatier, Paul, A. (1993). "Policy Change over a Decade or More." In Paul A. Sabatier and Hank C. Jenkins-Smith (eds.), *Policy Change and Learning: An Advocacy Coalition Approach*. Boulder, CO: Westview Press, pp. 13–40.

Sabatier, Paul A. and Hank C. Jenkins-Smith (eds.) (1993). *Policy Change and*

Learning: An Advocacy Coalition Approach. Boulder, CO: Westview Press, pp. 41–56.

Sachs, Jeffrey (1993). *Poland's Jump to the Market Economy*. Cambridge, MA: MIT Press.

Schmidt, Klaus M. and Monica Schnitzer (1993). "Privatization and Management Incentives in the Transition Period in Eastern Europe." *Journal of Comparative Economics* 17 (2): 264–287.

Schnoor, Jerald L., James N. Galloway and Bedrich Moldan (1997). "East Central Europe: An Environment in Transition." *Environmental Science and Technology* 31 (9): 412A–416A.

Schot, Johan W. (1992). "Constructive Technology Assessment and Technology Dynamics: The Case of Clean Technologies." *Science, Technology and Human Values* 17 (1): 36–56.

Schwarz, Michiel and Michael Thompson (1990). *Divided We Stand: Redefining Politics, Technology and Social Choice*. Philadelphia: University of Pennsylvania Press.

Seika, M., N. Metz and R. M. Harrison (1996). "Characteristics of Urban and State Emission Inventories—A Comparison of Examples from Europe and the United States." *The Science of the Total Environment* 189–190: 221–234.

Selden, Thomas M. and Daqing Song (1994). "Environmental Quality and Development: Is There a Kuznets Curve for Air Pollution Emissions?" *Journal of Environmental Economics and Management* 27 (2): 147–162.

Shleifer, Andrei (1997). *Government in Transition*. Development Discussion Paper No. 573. Cambridge, MA: Harvard Institute for International Development, Harvard University.

Shrader-Frechette, Kristin S. (1991). *Risk and Rationality: Philosophical Foundations for Populist Reforms*. Berkeley: University of California Press.

Shrivastava, Paul (1993). "The Greening of Business." In Denis Smith (ed.), *Business and the New Environment: Implications of the New Environmentalism*. New York: St. Martin's Press, pp. 27–39.

Silesian Scientific Institute (1989). *Social Issues in the Highly Industrial Regions*. Katowice: Published by the Silesian Scientific Institute (in Polish).

Skea, Jim and Steve Sorrell (eds.) (1999). *Pollution for Sale: Emissions Trading and Joint Implementation*. Cheltenham, England, and Northampton, MA: Edward Elgar.

Sleszyński, Jerzy (1998). "Economic Instruments in Polish Environmental Policy." In John Clark and Daniel H. Cole (eds.), *Environmental Protection in Transition: Economic, Legal and Socioeconomic Perspectives on Poland*. Brookfield, VT: Ashgate.

Slovic, Paul (1992). "Perception of Risk: Reflections on the Psychometric Paradigm." In Sheldon Krimsky and Dominic Golding (eds.), *Social Theories of Risk*. Westport, CT: Praeger Publishers, pp. 117–152.

Slovic, Paul and Robin Gregory (1999). "Risk Analysis, Decision Analysis, and the Social Context for Risk Decision Making." In Ward Edwards, James Shanteau, Barbara A. Mellers and David Schurn (eds.), *Decision Science and Technology: Reflections on the Contributions of Ward Edwards*. Norwell, MA: Kluwer Academic.

Smart, Bruce (ed.) (1992). *Beyond Compliance: A New Industry View of the Environment*. Washington, DC: World Resources Institute.

Smith, Denis (1993). "Business and the Environment: Towards a Paradigm Shift?" In Denis Smith (ed.), *Business and the Environment: Implications of the New Environmentalism*. New York: St. Martin's Press, pp. 1–11.

Smith, Leonard G. (1994). "Achieving Environmental Goals: The Concept and Practice of Environmental Performance Review." *Environment and Planning A* 26 (1): 157–158.

Smith, Mark and Stephen Potter (1996). "Eco-labelling and Environmental Policy: Policy Confusion Persists." *Policy Studies* 17 (1): 73–80.

Smolar, Aleksander (1996). "Revolutionary Spectacle and Peaceful Transition." *Social Research* 63 (2): 439–464.

Sorsa, Piritta (1994). *Competitiveness and Environmental Standards: Some Exploratory Results*. World Bank Policy Research Working Paper 1249. Washington, DC: World Bank.

Spiegel, Jerry and Annalee Yassi (1997). "The Use of Health Indicators in Environmental Assessment." *Journal of Medical Systems* 21 (5): 275–289.

State Labor Inspectorate (1995–1998). Annual Reports by the Chief Labor Inspector on *The Activities of the State Labor Inspectorate*, Warsaw (in Polish).

State Labor Inspectorate (1998). Report by the Chief Labor Inspector on the Activities of the State Labor Inspectorate, Warsaw (in Polish).

Steinzor, Rena I. (July 1998a). "Reinventing Environmental Regulation: Back to the Past by Way of the Future." *Environmental Law Reporter* 28: 10361–10363.

Steinzor, Rena I. (1998b). "Reinventing Environmental Regulation: The Dangerous Journey from Command to Self-Control." *The Harvard Environmental Law Review* 22 (1): 103–202.

Stern, Paul C. and Harvey V. Fineberg (eds.) (1996). *Understanding Risk: Informing Decisions in a Democratic Society*. Washington, DC: National Academy Press.

Stewart, Richard B. (1993). "Environmental Regulation and International Competitiveness." *Yale Law Journal* 102 (8): 2039–2106.

Stodulski, Wojciech (1996). *The International Monetary Fund and Sustainable Development*. Research Report #2. Warsaw: Institute for Sustainable Development.

Stodulski, Wojciech (May 1999). *10 Years of Transformation in Poland: Environmental Protection*. Warsaw: Institute for Sustainable Development (in Polish).

Szablowski, George J. (1993a). "East European Transitions, Elites, Bureaucracies, and the European Community." *Governance* 6 (3): 304–324.

Szablowski, George J. (1993b). "Governing and Competing Elites in Poland." *Governance* 6 (3): 341–357.

Sztompka, Piotr (1992). *Dilemmas of the Great Transition*. The Paul Nitze School of Advanced International Studies, Bologna Center Occasional Paper #74. Bologna, Italy: John Hopkins University Press.

Taras, Raymond (1995). *Consolidating Democracy in Poland*. Boulder, CO: Westview Press.

Tarkowska, Elżbieta (1993). "Culture and Uncertainty: A Few Remarks Concerning the Cultural Dimension of Transformation." *Sisiphus, Social Studies*. Warsaw: NFiS Publishers.

Tarkowska, Elżbieta and Jacek Tarkowski (Fall 1989). "Social Disintegration in Poland: Civil Society or Amoral Familism?" *Telos* 89: 103–109.

Tatur, Melanie (1995). "Social Movements and Institutional Change in Poland and Russia." *Research in Social Movements, Conflict and Change* 18: 19–35.

Thurmaier, Kurt (Winter 1994). "The Evolution of Local Government Budgeting in Poland: From Accounting to Policy in a Leap and a Bound." *Public Budgeting and Finance*: 84–97.

Tietenberg, Tom and David Wheeler (1998). "Empowering the Community: Information Strategies for Pollution Control." *Frontiers of Environmental Economics Conference*. October 23–25, Arlie House, Virginia.

Timberlake, L. (1981). "Poland: The Most Polluted Country in the World?" *New Scientist* 92 (1276): 248–250.

Toman, Michael A. (Fall 1993). "Using Economic Incentives to Reduce Air Pollution Emissions in Central and Eastern Europe: The Case of Poland." *Resources* 113: 18–23.

Toman, Michael A., Janusz Cofaba and Robin Bates (1994). "Alternative Standards and Instruments for Air Pollution Control in Poland." *Environmental and Resource Economics* 4 (5): 401–417.

Turner, Graham W., Rod M. C. Ruffio and Mark W. Roberts (1997). "Comparing Environmental Conditions Using Indicators of Pollution Hazard." *Environmental Management* 21 (4): 623–634.

Tymowski, Andrzej W. (1993). "Poland's Unwanted Social Revolution." *East European Politics and Societies* 7 (2): 169–202.

U.S. Department of Commerce (June 1998). National Trade Data Bank. *Poland: Economic Trends and Outlook*. Washington, DC: U.S. GPO.

U.S. EPA (1996). *1994 Toxics Release Inventory. Public Data Release*. EPA 745-R-96-002. Office of Pollution Prevention and Toxics. Washington, DC: U.S. EPA.

U.S. EPA (1997). *National Air Pollutant Trends 1900–1996*. EPA 464/R-97-001. Office of Air Quality Planning and Standards Research. Washington, DC: U.S. EPA.

Veneziano, Francesca (1997). "Access to Environmental Justice in a Country Applying for Membership of the EU: The Case of Poland." *European Business Law Review* 8 (5/6): 148–156.

Vig, Norman J. (1997). "Presidential Leadership and the Environment." In Norman Vig and Michael Kraft (eds.), *Environmental Policy in the 1990s: Reform or Reaction?* Third Edition. Washington, DC: CQ Press, pp. 95–118.

Vogel, David (1986). *National Styles of Regulation: Environmental Policy in Great Britain and the United States*. Ithaca NY: Cornell University Press.

von Winterfeldt, Detlof (1992). "Expert Knoweldge and Public Values in Risk Management: The Role of Decision Analysis." In Sheldon Krimsky and Dominic Golding (eds.), *Social Theories of Risk*. Westport, CT: Praeger Publishers, pp. 321–342.

Voorhees, John and Robert A. Woellner (1997). *International Environmental Risk Management: ISO 14000 and the Systems Approach*. Boca Raton, FL: CRC Press.

Wajda, Stanislaw and Jerzy Sommer (1994). "Environmental Liability in Property Transfer in Poland." In Gretta Goldenman (ed.), *Environmental Liability and Privatization in Central and Eastern Europe*. Boston: Kluwer Academic.

Walicki, Andrzej (1990). "Three Traditions in Polish Patriotism." In Stanisław

Gomułka and Antony Polonśky (eds.), *Polish Paradoxes*. London: Routledge.
Walicki, Andrzej (1997). "Transitional Justice and the Political Struggles of Post-Communist Poland." In James A. McAdams (ed.), *Transitional Justice and the Rule of Law in New Democracies*. Notre Dame and London: University of Notre Dame Press, pp. 185–237.
Waller, Michael and Frances Millard (1992). "Environmental Politics in Eastern Europe." *Environmental Politics* 1: 159–185.
Wargo, John (1996). *Our Children's Toxic Legacy: How Science and Law Fail to Protect Us from Pesticides*. Second Edition. New Haven CT: Yale University Press.
Warner, Jonathan (1996). "The Environment in Poland." In Zenon Wisniewski (ed.), *The Adjustment of Polish Companies to the Market Economy*. Torun, Poland: Mikolaj Kopernicus University Press, pp. 79–101.
Weale, Albert (1995). "The Kaleidoscopic Competition of European Environmental Regulation." *European Business Journal* 7 (4): 19–25.
Weber, Edward P. (1998a). *Pluralism by the Rules: Conflict and Cooperation in Environmental Regulation*. Washington, DC: Georgetown University Press.
Weber, Edward P. (1998b). "Successful Collaboration: Negotiating Effective Regulations." *Environment* 40 (9): 10–15, 32–37.
Webler, Thomas (1995). " 'Right' Discourse in Citizen Participation: An Evaluative Yardstick." In Ortwin Renn, Thomas Webler and Peter M. Wiedemann (eds.), *Fairness and Competence in Citizen Participation: Evaluating Models for Environmental Discourse*. Boston: Kluwer Academic, pp. 35–86.
Webler, Thomas (1997). "Organizing Public Participation: A Review of Three Handbooks." *Human Ecology Review* 3 (1): 345–354.
Wedel, Janine R. (1988). *The Private Poland*. New York: Facts on File.
Wedel, Janine R. (ed.) (1992). *The Unplanned Society: Poland during and after Communism*. New York: Columbia University Press.
Welch, Thomas E. (1998). *Moving Beyond Environmental Compliance: A Handbook for Integrating Pollution Prevention with ISO 14000*. Boca Raton, FL: CRC Lewis Publishers.
Welford, Richard (ed.) (1996). *Corporate Environmental Management: Systems and Strategies*. London: Earthscan Publications.
Welford, Richard and Richard Starkey (eds.) (1996). *Business and the Environment: A Reader*. Washington, DC: Taylor and Francis Publishers.
Wever, Grace (1996). *Strategic Environmental Management: Using TQEM and ISO 14000 for Competitive Advantage*. New York: John Wiley and Sons.
White, Allen L. (1999). "Sustainability and the Accountable Corporation." *Environment* 41 (8): 3–43.
White, Keith A., Joyce K. Smith and John L. Warren (1994). "Developing a Decision Support Tool for Life-Cycle Cost Assessments." *Total Quality Environmental Management* 4 (1): 23–26.
Wildavsky, Aaron B. (1987). "Choosing Preferences by Constructing Institutions: A Cultural Theory of Preference Formation." *American Political Science Review* 81 (March): 2–21.
Wildavsky, Aaron B. (1988). *Searching for Safety*. New Brunswick, NJ: Transaction Books.

Wildavsky, Aaron B., Richard J. Ellis and Michael Thompson (eds.) (1997). *Culture Matters: Essays in Honor of Aaron Wildavsky*. Boulder, CO: Westview Press.

Williams, Bernard (1988). "Formal Structures and Social Reality." In Deigo Gambetta (ed.), *Trust: Making a Breaking Cooperative Relations*. Oxford: Blackwell, p. 8.

Winiecki, Jan (1997). *Institutional Barriers to Poland's Economic Development*. New York: Routledge.

Wiśniewski, Zenon (ed.) (1996). *The Adjustment of Polish Companies to the Market Economy*. Torun, Poland: Uniwersytet Nikolaja Kopernika.

Wokutch, Richard E. (1990). *Cooperation and Conflict in Occupational Safety and Health: A Multination Study of the Automotive Industry*. New York: Praeger Publishers.

Wollmann, H. (1997). "Institution Building and Decentralization in Formerly Socialist Countries: The Case of Poland, Hungary, and East Germany." *Environment and Planning C: Government and Policy* 15: 463–480.

World Bank (1999). "Greening Industry." http://www.worldbank.org/publications/ordform.greenord.

World Bank (2000). *Greening Industry: New Roles for Communities, Markets, and Governments*. A World Bank Policy Research Report. New York: Oxford University Press.

World Health Organization (1996). *Health for All*. Geneva: WHO Regional Office for Europe.

World Resources Institute (1992). *World Resources 1991–93: A Guide to the Global Environment: Towards Sustainable Development*. New York: Oxford University Press.

World Resources Institute (1998). *World Resources 1997*. Washington, DC: World Resources Institutes.

Wright, R.A.D. (1994). "Environmental Management Systems: A Quantam Leap Forward or Nothing New for the E&P Industry." *The APEA Journal* 34 (1): 771.

Yosie, Terry F. and Timothy D. Herbst (1998). *Using Stakeholder Processes in Environmental Decisionmaking: An Evaluation of Lessons Learned, Key Issues, and Future Challenges*. Washington, DC: American Industrial Health Council.

Young, C. William (1996). "Measuring Environmental Performance." In Richard Welford (ed.), *Corporate Environmental Management: Systems and Strategies*. London: Earthscan Publications, pp. 150–176.

Zamoyski, Adam (1987). *The Polish Way: A Thousand-Year History of the Poles and Their Culture*. London: Murray.

Zechenter, Elżbieta M. (1993). "The Socio-economic Transformation of Poland: Privatization and the Future of Environmental Protection." *Georgetown International Environmental Law Review* 6 (1): 99–149.

Zuzowski, Robert (1993). "Political Culture and Dissent: Why Were There Organizations Like KOR in Poland?" *East European Quarterly* 27 (4): 48–72.

Żylicz, Tomasz (1994). "In Poland, It's Time for Economics." *Environmental Impact Assessment Review* 14 (2/3): 79.

Index

About the Authors

HALINA SZEJNWALD BROWN is Professor of Environmental Health at Clark University. She has written widely on the use of science in public policy, health and safety issues in industrial enterprises, corporate hazard management and environmental health policy.

DAVID ANGEL holds the Leo L. and Joan Laskoff Professorship in Economics, Technology, and the Environment at Clark University. His specialty is economic geography, and his research interests focus on issues of industrial restructuring and technological change.

PATRICK G. DERR is Professor of Philosophy and Director of the Program in Ethics and Public Policy at Clark University. He is a member of the Hazards Research Group at the Center for Technology, Environment and Development and writes on occupational and environmental hazards, history and philosophy of science and health care.